The Handbook of Environmental Chemistry

Founded by Otto Hutzinger

Editors-in-Chief: Damià Barceló • Andrey G. Kostianoy

Volume 9

The Handbook of Environmental Chemistry
Recently Published and Forthcoming Volumes

Polymers – Opportunities and Risks II: Sustainability, Product Design and Processing
Volume Editors: P. Eyerer, M. Weller, and C. Hübner
Vol. 12, 2010

Polymers – Opportunities and Risks I: General and Environmental Aspects
Volume Editor: P. Eyerer
Vol. 11, 2010

Chlorinated Paraffins
Volume Editor: J. de Boer
Vol. 10, 2010

Biodegradation of Azo Dyes
Volume Editor: H. Atacag Erkurt
Vol. 9, 2010

Water Scarcity in the Mediterranean: Perspectives Under Global Change
Volume Editors: S. Sabater and D. Barceló
Vol. 8, 2010

The Aral Sea Environment
Volume Editors: A.G. Kostianoy and A.N. Kosarev
Vol. 7, 2010

Alpine Waters
Volume Editor: U. Bundi
Vol. 6, 2010

Transformation Products of Synthetic Chemicals in the Environment
Volume Editor: A.B.A. Boxall
Vol. 2/P, 2009

Contaminated Sediments
Volume Editors: T.A. Kassim and D. Barceló
Vol. 5/T, 2009

Biosensors for the Environmental Monitoring of Aquatic Systems
Bioanalytical and Chemical Methods for Endocrine Disruptors
Volume Editors: D. Barceló and P.-D. Hansen
Vol. 5/J, 2009

Environmental Consequences of War and Aftermath
Volume Editors: T.A. Kassim and D. Barceló
Vol. 3/U, 2009

The Black Sea Environment
Volume Editors: A. Kostianoy and A. Kosarev
Vol. 5/Q, 2008

Emerging Contaminants from Industrial and Municipal Waste
Removal Technologies
Volume Editors: D. Barceló and M. Petrovic
Vol. 5/S/2, 2008

Fuel Oxygenates
Volume Editor: D. Barceló
Vol. 5/R, 2007

The Rhine
Volume Editor: T. P. Knepper
Vol. 5/L, 2006

Persistent Organic Pollutants in the Great Lakes
Volume Editor: R. A. Hites
Vol. 5/N, 2006

Antifouling Paint Biocides
Volume Editor: I. Konstantinou
Vol. 5/O, 2006

Estuaries
Volume Editor: P. J. Wangersky
Vol. 5/H, 2006

The Handbook of Environmental Chemistry Also Available Electronically

The Handbook of Environmental Chemistry is included in Springer's eBook package *Earth and Environmental Science*. If a library does not opt for the whole package, the book series may be bought on a subscription basis.

For all customers who have a standing order to the print version of *The Handbook of Environmental Chemistry*, we offer the electronic version via Springer Link free of charge. If you do not have access, you can still view the table of contents of each volume and the abstract of each article on SpringerLink (www.springerlink.com/content/110354/).

You will find information about the

- Editorial Board
- Aims and Scope
- Instructions for Authors
- Sample Contribution

at springer.com (www.springer.com/series/698).

All figures submitted in color are published in full color in the electronic version on SpringerLink.

Aims and Scope

Since 1980, *The Handbook of Environmental Chemistry* has provided sound and solid knowledge about environmental topics from a chemical perspective. Presenting a wide spectrum of viewpoints and approaches, the series now covers topics such as local and global changes of natural environment and climate; anthropogenic impact on the environment; water, air and soil pollution; remediation and waste characterization; environmental contaminants; biogeochemistry; geoecology; chemical reactions and processes; chemical and biological transformations as well as physical transport of chemicals in the environment; or environmental modeling. A particular focus of the series lies on methodological advances in environmental analytical chemistry.

Series Preface

With remarkable vision, Prof. Otto Hutzinger initiated *The Handbook of Environmental Chemistry* in 1980 and became the founding Editor-in-Chief. At that time, environmental chemistry was an emerging field, aiming at a complete description of the Earth's environment, encompassing the physical, chemical, biological, and geological transformations of chemical substances occurring on a local as well as a global scale. Environmental chemistry was intended to provide an account of the impact of man's activities on the natural environment by describing observed changes.

While a considerable amount of knowledge has been accumulated over the last three decades, as reflected in the more than 70 volumes of *The Handbook of Environmental Chemistry,* there are still many scientific and policy challenges ahead due to the complexity and interdisciplinary nature of the field. The series will therefore continue to provide compilations of current knowledge. Contributions are written by leading experts with practical experience in their fields. *The Handbook of Environmental Chemistry* grows with the increases in our scientific understanding, and provides a valuable source not only for scientists but also for environmental managers and decision-makers. Today, the series covers a broad range of environmental topics from a chemical perspective, including methodological advances in environmental analytical chemistry.

In recent years, there has been a growing tendency to include subject matter of societal relevance in the broad view of environmental chemistry. Topics include life cycle analysis, environmental management, sustainable development, and socio-economic, legal and even political problems, among others. While these topics are of great importance for the development and acceptance of *The Handbook of Environmental Chemistry,* the publisher and Editors-in-Chief have decided to keep the handbook essentially a source of information on "hard sciences" with a particular emphasis on chemistry, but also covering biology, geology, hydrology and engineering as applied to environmental sciences.

The volumes of the series are written at an advanced level, addressing the needs of both researchers and graduate students, as well as of people outside the field of "pure" chemistry, including those in industry, business, government, research establishments, and public interest groups. It would be very satisfying to see these volumes used as a basis for graduate courses in environmental chemistry. With its high standards of scientific quality and clarity, *The Handbook of*

Environmental Chemistry provides a solid basis from which scientists can share their knowledge on the different aspects of environmental problems, presenting a wide spectrum of viewpoints and approaches.

The Handbook of Environmental Chemistry is available both in print and online via www.springerlink.com/content/110354/. Articles are published online as soon as they have been approved for publication. Authors, Volume Editors and Editors-in-Chief are rewarded by the broad acceptance of *The Handbook of Environmental Chemistry* by the scientific community, from whom suggestions for new topics to the Editors-in-Chief are always very welcome.

Damià Barceló
Andrey G. Kostianoy
Editors-in-Chief

Volume Preface

Synthetic dyes are extensively used in textile, dyeing, paper, printing, color photography, pharmaceutical, food, cosmetics, and other industries. Azo dyes represent about one-half of all the dyes in common use and are employed as coloring agents in the textile, food, and pharmaceutical industries. Disposal of waste water from textile industries is a problem in many parts of the world. Although these dyes are not toxic in themselves, after being released into the aquatic environment, they may be converted into potentially carcinogenic amines that impact the ecosystem and human health. The absorption of light due to textile dyes creates problems to photosynthetic aquatic plants and algae. Nowadays, the public demand for colour-free discharges to receiving water bodies has made decolourisation of a variety of industrial waste water a top priority.

Chemical and physical methods including adsorption, coagulation-flocculation, advanced oxidation and electrochemical methods are very efficient in color removal. These methods are quite expensive and have operational problems. High sludge formation, regeneration requirement and cost of adsorbent make adsorption an unattractive method for decolorization purposes. So biodegradation begins to play an important role in decolorization of azo dyes.

This volume of The Handbook of Environmental Chemistry is very important as it includes different biodegradation methods with different microorganism groups. Integration of biological processes with physical and chemical processes are also given in this volume. Several biodegradation methods can be found in this one book and it is possible to compare these methods. All the chapters in this volume have been written by authors who are experts in the field.

This book is divided into 11 chapters. The first chapter outlines the bioaugmentation of azo dyes, a process in which various microorganisms are applied to the bioreactor or the polluted sites to accelerate the desired biological processes. The second chapter focuses on the different anaerobic microbial processes of biodegradation of azo dyes and enzymes that are responsible for their degradation. The third chapter reviews the biodegradation of azo dyes in anaerobic-aerobic sequencing batch reactors, where the cyclic operations of SBR provide both color removal in the anaerobic stage and aromatic amine removal in the aerobic stage. The fourth chapter outlines azo dye degradation by immobilized bacteria and concludes that immobilization increases the stability of the enzyme at high pH and tolerance to elevated temperatures and makes the enzyme less vulnerable to inhibitors. The fifth chapter focuses on bacterial decolorization and degradation of azo dyes catalyzed by redox mediators and the further investigation to enhance the applicability of redox mediators to the bio-transformation of azo dyes. In the sixth chapter, a survey of the state-of-the-art of azo-dye conversion by means of bacteria is presented with a focus on reactor design and operational issues. The relevance of thorough characterization of reaction kinetics and yields is discussed. The second section focuses on recent results regarding the conversion of an azo-dye by means of bacterial biofilm in an internal loop airlift reactor. Experimental results are analyzed in the light of a comprehensive reactor model. The seventh chapter outlines the treatment of azo dye-containing waste water using integrated processes like combined physical biological processes and combined chemical biological processes. The eighth chapter is about the role of white rot fungi in biodegradation of azo dyes and the detection of enzymes responsible for azo dye decolorization. The ninth chapter is about decolorization of azo dyes by immobilized fungi. The tenth chapter focuses on decolorization of azo dyes with another fungus group: yeasts.

The last chapter highlights the factors affecting the complete mineralization of azo dyes.

Nicosia, North Cyprus
February 2010

Hatice Atacag Erkurt
Volume Editor

Contents

Bioaugmentation of Azo Dyes

Azeem Khalid, Muhammad Arshad, and David Crowley

Abstract Biodegradation is a cost-effective method to remove the residues of azo dyes prior to their discharge in wastewater streams from dye product industries. The efficacy of this treatment method is highly dependent on establishing an effective degrader community and maintaining environmental conditions that support the growth and activity of the degrader organisms. Although activated sludge is commonly used as a source of degrader organisms to start the process, bioaugmentation of the wastewater with highly effective strains provides a much more reliable process in which the process manager can use bacterial strains that target particular dye chemicals and metabolites to achieve complete mineralization. The most effective inoculants are able to degrade dyes over a broad concentration range, tolerate a range of environmental conditions of temperature, pH, and salinity, and persist at high population densities in competition with other microorganisms in mixed microbial cultures. The use of growth supplements such as yeast extract can further enhance the biodegradation activity. The ability to achieve complete mineralization of azo dyes depends on the control of the process in which initial decolorization takes place under microaerophilic conditions with low oxygen, followed by elimination of the dye metabolites using an aeration step. In many cases, this may be best achieved by using a mixture of bacterial strains that sequentially carry out the two-step process. Practical development of bacteria for bioaugmentation requires careful screening that is based not only on their efficacy in pure culture, but also on their ability to compete with the indigenous microbial communities in wastewater streams and ability to be produced and delivered as

A. Khalid
Department of Environmental Sciences, PMAS Arid Agriculture University, Rawalpindi, Pakistan

M. Arshad
Institute of Soil and Environmental Sciences, University of Agriculture, Faisalabad 38040, Pakistan

D. Crowley (✉)
Department of Environmental Sciences, University of California, Riverside, CA 92521, USA
e-mail: crowley@ucr.edu

H. Atacag Erkurt (ed.), *Biodegradation of Azo Dyes*,
Hdb Env Chem (2010) 9: 1–37, DOI 10.1007/698_2009_42,

a stable inoculum. In the future, it may be useful to consider bioaugmentation with bacteria that contain mobile genetic elements that carry catabolic pathways, thereby allowing the genes to be introduced into the indigenous microorganisms. The ability to monitor introduced bacteria or catabolic genes will continue to be important for process optimization both in the laboratory and during operation in full-scale treatment systems.

Keywords Bioaugmentation, Bioreactor, Environmental factors, Inoculation, Salinity

Contents

Abbreviations

AR	Acid red
BOD	Biological oxygen demand
DBMR	Direct brown MR
DO	Disperse orange
DR	Direct red
MGE	Mobile genetic element
NAD(P)H	Nicotinamide adenine dinucleotide (phosphate)
RAPD	Randomly amplified polymorphic DNA
RB	Reactive black

1 Introduction

Treatment of dye-contaminated wastewater discharged from the textile and other dye-stuff industries is necessary to prevent contamination of soil and surface and ground water. Currently, there are several physicochemical and biological methods for the removal of dyes from effluents [1–12]. Among these, biotechnological approaches are receiving increased attention worldwide as environmental-friendly methods that are becoming increasingly efficient and cost-effective for the remediation of dye-contaminated wastewater [13, 14]. Many biotreatment systems rely on the use of sludge as an inoculum to initiate the dye degradation process [15–18]. While generally effective, it is nonetheless important to assure complete mineralization and detoxification for use as a reliable treatment method. Azo dyes and their degradation intermediates vary in their recalcitrance to biodegradation due to their compiex structures and xenobiotic nature and in some cases are both mutagenic and carcinogenic [19–26]. Furthermore, azo-dye degrading microbial communities are sensitive to high concentrations of salts that are used in the dye process [27, 28]. This can limit growth and activity of the degrader bacteria such that the process treatment times become impractical. With the discovery and isolation of very efficient, salt-tolerant azo-dye degrading bacteria, bioaugmentation of biotreatment systems with specific microbial strains has now become an effective strategy to improve wastewater treatment systems and to enhance the bioremediation of azo dyes [29–33].

Bioaugmentation is a process in which various microorganisms including indigenous, wild type, or genetically engineered are introduced to the bioreactor or the polluted sites/matrices to accelerate the desired biological processes and achieve more consistent results [31, 34]. As used here, bioaugmentation refers to the use of selected strains of bacteria as opposed to the use of nonspecific microbial cultures such as those that are contained in activated sludge. Although activated sludge is used in the process of treating the contaminants, the microbial species that are contained in this material are uncharacterized and the system is a "black box" [35]. This may lead to inconsistent results, such that in some cases, up to 90% of the dyes in an effluent can remain untreated after an activated sludge process [36]. On the other hand, activated sludge can provide a useful starting medium from which individual strains or consortia can be isolated and cultured for use as inoculants [14, 37–41]. The microbial species and consortia can then be studied to determine the environmental factors that affect their growth and the rate of degradation.

Often the effectiveness of individual isolates can be enhanced by co-culture with other highly efficient dye-decolorizing strains [42–44]. Here, it is speculated that the combined enzyme systems of the mixed bacterial culture are more effective than the enzymes from the individual isolates, each of which may have different substrate kinetics and efficiency at different dye concentrations. Cooperation within microbial communities also can occur through exchange of growth cofactors and removal of toxic metabolites. Although many microorganisms can degrade azo dyes [7, 40–42, 45–51], relatively few microbial species and strains have emerged as candidates for use in bioaugmentation [14, 40, 52–55]. Before individual isolates can be recommended, comprehensive research is required to understand the role of individual microorganisms and their interactions with other microflora [24, 35, 56]. In this chapter, various types of azo dye degrading microorganisms and their potential for bioaugmentation are discussed.

2 Azo Dye Degrading Bacteria

2.1 *Isolation of Azo Dye Degrading Bacteria*

Several studies have demonstrated partial or complete degradation of dyes by pure and mixed cultures of bacteria (Table 1). In many biotreatment systems, mixed bacterial cultures have proved to be superior to single pure cultures. It has been reported that a higher degree of azo dye biodegradation might be achieved by mixed bacterial cultures due to complementary catabolic pathways within the microbial community that may not be accomplished by individual pure strains [48, 49, 76, 88–90]. Recently, however, several researchers have identified single bacterial strains that have very high efficacy for removal of azo dyes [14, 40, 41, 84, 86, 91]. In contrast to mixed cultures, the use of a pure culture has several advantages. These include predictable performance and detailed knowledge on the degradation pathways with improved assurance that catabolism of the dyes will lead to nontoxic end products under a given set of environmental conditions. Another advantage is that the bacterial strains and their activity can be monitored using culture-based or molecular methods to quantify population densities of the bacteria over time. Knowledge of the population density can be extrapolated to quantitative analysis of the kinetics of azo-dye decoloration and mineralization.

2.2 *Redox Control of the Degradation Process*

Biodegradation of dyes can be achieved under both aerobic and anaerobic conditions, but involves different metabolic pathways that affect the process rates and metabolites that are produced from the parent chemicals. Aerobic treatment is one of the most commonly used treatment methods for wastewater, but is often less effective for facilitating degradation of dyes than an alternating anaerobic–aerobic treatment or microaerophilic treatment system. Initial decolorization of azo dyes is known to involve a reductive process (Fig. 1) and is thus facilitated by anaerobic, static culture conditions [40, 48, 49, 90, 92–101]. The intracellular or extracellular process by which the dyes are reduced is not yet clear. In vitro, two types of NAD (P)H-dependent cytoplasmic azo-reductases have been described by Chen [102], but significant reductase activity is observed only with cell extracts, as opposed to incubation with intact bacterial cells [103, 104]. Since many dyes are polar and/or are large molecules for which there may not be carrier proteins, it is unlikely that they may pass through the cell membrane to enter the interior of the cell where they can be utilized by nonspecific reductase enzymes. Therefore, it is hypothesized that bacterial dye reduction is mainly an extracellular process [10, 105–107]. This hypothesis is supported by the findings that microbial excreted or artificial redox mediators catalyze the dye decolorization process [105, 106, 108].

Table 1 Bacterial species capable of degrading azo dyes

Bacterial species	Dyes	Comments	References
Acinetobacter calcoaceticus NCIM 2890	Direct brown MR (DBMR)	Decolorization of DBMR was 91.3% in static anoxic condition, whereas agitated cultures showed less decolorization (59.3%) after 48 h	[57]
Acinotobacter sp., *Citrobacter freundii*, *Klebsiella oxytoca*	Acid Red 88, Reactive Black 5, Direct Red 81, Disperse Orange 3	The mixed culture of bacteria removed 88–100% dyes (100 mg L^{-1}) in 10 h	[14]
Aeromonas caviae, *Proteus mirabilis*, *Rhodococcus* sp.	Acid Orange 7	More than 90% decolorization of the dye was achieved in 16 h	[58]
Aeromonas, *Pseudomonas*, *Bacillus*, *Shewanella and Massillia* spp.	Reactive Black 5, Direct Red 81, Acid Red 88, Disperse Orange 3	Treatment times required by the most efficient strain, AS96 (*Shewanella putrefaciens*), were as short as 4 h for complete decolorization of 100 mg L^{-1} of AR-88 and DR-81 dyes under static conditions, and 6 and 8 h, respectively, for complete decolorization of RB-5 and DO-3	[40, 41]
Bacillus cereus DC11	Acid Blue 25, Malachite Green, Basic Blue	High decolorization efficiency (95–98%) achieved within 6 h for 100 mM Acid Blue 25 (anthraquinone dye), 4 h for 55 mM Malachite Green (triphenylmethane dye), and 2 h for 750 mM Basic Blue X-GRRL under anaerobic conditions	[59]
Bacillus fusiformis	Disperse Blue 79, Acid Orange 10	The dyes were completely mineralized within 48 h	[60]
Bacillus sp.	Congored	The complete decolorization was achieved in 24–27 h for a concentration of 100–300 mg dye L^{-1}	[61]
Bacillus subtilis HM	Fast Red	Under the near-optimal conditions, 99% of the decolorization was achieved in 6 h	[62]

(*continued*)

Table 1 (continued)

Bacterial species	Dyes	Comments	References
Bacillus thurengiensis	Acid Red 119	The dye was decolorized up to 70% in 24 h	[63]
Bacillus velezensis AB	Direct Red 28	The dye (25 mg L^{-1}) was completely decolorized within 10 h.	[64]
Citrobacter sp. CK3	Reactive Red 180	About 95% dye (200 mg L^{-1}) was removed in 36 h	[65]
Enterococcus gallinarum	Direct Black 38	The bacterium removed 53–63% of the dye in 24 h in minimal medium while 71–85% of decolourization was observed in Luria broth medium.	[66]
Escherichia coli NO3	Reactive Red 22	After acclimation, time for 50% color removal lowered from 5.7 to 4.3 h	[67]
Escherichia coli, Pseudomonas sp.	Congo Red, Direct Black 38	The complete decolorization was achieved at the end of 9 days of incubation in case of *E. coli* while *Pseudomonas* sp. decolorized in 5 days	[68]
Eschericia coli YB	Acid Red 27	The dye was decolorized up to 75% in 2 h	[69]
Halomonas sp.	Reactive Brilliant Red X, Acid Black 10B, Acid Scarlet GR, Acid Red B, Acid Red G, Reactive Brilliant Red K	The decolorization of the dyes was up to 90% in 24 h	[70–72]
Halomonas sp.	Remazol Black, Maxilon Blue, Sulfonyl Scarlet BNLE, Sulfonyl Green BLE, Remazol Black N, Entrazol Blue IBC	The bacterium was capable of decolorizing the dyes in wide range of NaCl concentrations after 4 days of incubation period	[73]
Kerstersia sp. VKY1	Amaranth, Fast R, Ponceau S, Congo R, Orange II, Acid O 12, Acid R 151	The first four dyes decolorized by the bacterium by 100% while the remaining three decolorized by 84, 73 and 44%, respectively, in 24 h	[74]

(*continued*)

Table 1 (continued)

Bacterial species	Dyes	Comments	References
Klabisiella sp. VN-31	Reactive Yellow 107, Reactive Red 198, Reactive Black 5, Direct Blue 71	Monoazo dyes RY107 and RR 198 were decolorized in 72 and 96 h; the diazo dyes (RB5 and triazodye DB71) decolorized in 120 and 168 h	[39]
Lactobacillus casei TISTR 1500	Methyl Orange	The complete decolorization of the dye was achieved in 2.5 h	[75]
Paenibacillus polymyxa, Micrococcus luteus	Reactive Violet 5R	The bacterial consortium showed complete decolorization in 36 h	[76]
Proteus vulgaris, Micrococcus glutamicus	Scarlet R	Bacterial consortium decolorized 90% dye in 3 h	[77]
Pseudomonas luteola	Reactive azo dyes, Direct azo dyes and leather dyes	The 59–99% color removal after 2–6 days static incubation, at dye concentration of 100 mg L^{-1}, monoazo dyes showing fastest rate of decoloration	[78]
Pseudomonas aeruginosa, P. oleovarons, P. putida	Methyl Orange, Y87, B86, R91, B19, R90, B69, B31, B36, Y15, R34, B15, Y79, and B54	*P. aeruginosa* showed decolorization efficiency over 98% after 48 h while 76% decolorization was achieved by *P. oleovarons* after 54 h. *P. putida* showed lower efficiency	[79]
Pseudomonas desmolyticum	Direct Blue 6, Green HE4B, Red HE7B	The dye GHE4B was completely decolorized in 12 h while DB 6 and RHE7B were decolorized in 16 h	[80]
Pseudomonas luteola, Eschericia coli	Reactive Red 22	The *E. coli* improved the ability of *Pseudomonas* sp. to decolorize the dye by producing decolorization – stimulating extracellular metabolites	[42]
Pseudomonas putida mt-2	Acid Violet 7	Complete biodegradation of azo dye up to 200 mg L^{-1} was achieved in 49 h under shaking while the biodegradation time was reduced to 37 h under static conditions	[81]

(*continued*)

Table 1 (continued)

Bacterial species	Dyes	Comments	References
Pseudomonas sp. SUK1	Reactive Red 2	The strain was capable of degrading dye in a wide range of concentration (up to 5 g L^{-1}) and almost 80% dye was removed in 114 h	[82]
Rhodopseudomonas palustris	Reactive Black 5	The dye up to 700 mg L^{-1} concentration was complete decolorized in 40 h	[83]
Shewanella decolorationis S12	Fast Acid Red GR	After 4 h incubation, more than 90% of the color was removed under anaerobic conditions while 12.8 and 33.7% decolorizing rates were observed under aerobic and microaerophilic conditions	[84]
Shewanella decolorationis sp. nov. S 12^{T}	Fast Acid Red GR, Reactive Brilliant Blue	The 90% decolorization of the dyes was achieved within 12 h	[85]
Shewanella J18 143	Remazol Black B, Acid Orange 7	Anaerobic cultures of *Shewanella* strain J18 143 rapidly removed color from the azo dye Remazol Black B in the growth medium to produce an absorbance at 597 nm of less than 1 in under 40 min	[86]
Sphingomonas herbicidovorans	Anthraquinone dyes	The bacterium was capable of decolorizing bromoamine acid dye (1,000 mg L^{-1}) more than 98% within 24 h	[87]

Recently, Brigé et al. [109] demonstrated that dye decolorization is an extracellular reduction process requiring a multicomponent electron transfer pathway that consists of cytoplasmic membrane, periplasmic, and outer membrane components. Similarly, we have demonstrated the ability of bacteria to remove the color of azo dyes from solid agar medium, which suggested the accumulation of redox active enzymes or biochemical substances that were released into the medium during growth of the bacterial cells [40]. These studies imply that reducing equivalents are transferred from an intracellular electron transport chain to the mediators, which consequently reduces the extracellular dye non-enzymatically. Another possibility is that the bacteria establish a link between their intracellular electron transport systems and the extracellular dye via electron transferring proteins in the outer

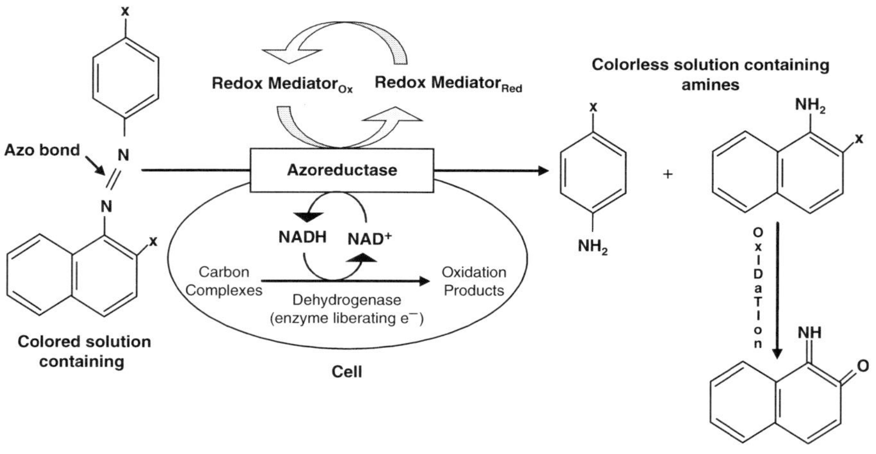

Fig. 1 Possible mechanisms for the removal of azo dyes by bacteria (modified from [86])

membrane [10, 109], ultimately reducing the dye either directly or indirectly via redox mediators.

2.3 *Cosubstrates*

Since the azo dye does not yield carbon or energy for growth during the first stage of enzymatic attack, various organic compounds (cosubstrates) are required for the dye decolorization step, in which the dyes act as acceptors of electrons that are supplied through the reducing equivalents that are generated by the electron transport chain [109]. Azo dye decolorization by mixed as well as pure cultures, generally, requires organic sources such as glucose, starch, acetate, ethanol, peptone yeast extract, or a combination of complex organic sources and carbohydrates [41, 42, 48, 49, 76, 84, 85, 95, 109]. As a result of the anaerobic reduction step, a variety of colorless aromatic compounds are synthesized. Depending on their chemical properties, these metabolites will accumulate under the anaerobic conditions, in which case further degradation can be achieved at accelerated rates under aerobic conditions [14, 86].

Previously, Kudlich et al. [110] reported that such types of compounds undergo rapid oxidation reactions, forming a range of polycyclic intermediates. In this manner, a sequential anaerobic–aerobic system is preferred for complete degradation of dyes [14, 41, 84]. The cleavage of azo linkages is not specific under anaerobic conditions [15, 111–115]; however, the electron withdrawing nature of the azo linkages impedes the susceptibility of dye molecules to oxidative reaction [116] and, thus, azo dyes show resistance to aerobic biodegradation [117–120]. Nonetheless, some bacteria with azo dyes-reducing enzymes, both specific and nonspecific, were capable of degrading azo dyes under aerobic conditions [40, 115, 121–125].

3 Substrate Specificity of Azoreductase for Different Types of Azo Dyes

Azo dyes are a diverse class of chemicals in which various moieties confer a wide range of colors. The number and position of sulfonate and other substituent groups on the azo dye are particular features that affect the rate of decolorization. Hitz et al. [126] illustrated that acid dyes exhibit low color removal due to the number of sulfonate groups present in the dye, while the direct dyes exhibit high levels of color removal, being independent of the sulfonate groups. As illustrated in studies with *Lactobacillus casei* TISTR 1500, methyl red with a mono-azo bond and lacking a sulfonate group is relatively easily degraded, while acid red 151 and congo red with two azo bonds are difficult to cleave [75]. Similarly, the decolorization rates observed in case of acid red and acid orange 8 were lower than those of other dyes containing sulfonate groups [75]. The resistance to degradation shown by the latter dyes could be attributed to their complicated chemical structures consisting of polyaromatic and sulfonate groups. This can be attributed to steric interference and increased difficulty for azoreductases to form enzyme substrate complexes with acid red 151 and acid orange 8. Likewise, dyes with methyl, methoxy, sulfo, or nitro groups in their structures and substituent groups in the molecule also affect azoreductase activity [10, 103, 127]. Nigam et al. [90] suggested that azo compounds with a hydroxyl group or with an amino group are more likely to be degraded at faster rates than those with a methyl, methoxy, sulfo, or nitro groups.

Zimmermann et al. [125] suggested some general structural features of dye substrates for reduction by azoreductases. They viewed that a hydroxy group in the *ortho* position of the naphthol ring is a prerequisite for the azoreductase reaction, and charged groups in the proximity of the azo group could cause hindrance in the reaction. Similarly, a second polar substituent on the dye molecule inhibits the reaction by lowering its affinity to the enzyme, while the electron withdrawing substituents on the phenyl ring increases the rate of the reaction. The dye reduction rates are also influenced by changes in electron density in the region of the azo group. The substitution of electron withdrawing groups in the *para* position of the phenyl ring, relative to the azo bond, causes an increase in the reduction rate [128]. Hydrogen bonding, in addition to the electron density in the region of the azo bond, has a significant effect on the rate of reduction [129]. It was also shown that sulfonated dyes were reduced faster than carboxylated dyes due to the higher electronegativity of the sulfo group, which renders the azo group more accessible to electrons [130]. Likewise, Martins et al. [131] reported that dyes with low polarity and having an electron-donating methyl substituent group in the ring are quite recalcitrant. Thus, it can be concluded that the decolorization of azo dyes is highly dependent on the specificity of azoreductase for different types of azo dyes that affect formation of substrate–enzyme complexes and the ability of the dye to accept an electron and cleave the azo group from the parent molecule.

4 Isolation, Enrichment, and Screening of Azo Dye Degrading Bacteria

Identification of azo dye degrading bacterial strains for use in bioaugmentation typically involves a stepwise process to isolate potential strains and screen them for their ability to degrade different dyes. A number of strategies have been devised to isolate such bacteria to achieve consistent and reproducible results in biotreatment systems (Fig. 2). Specific methods that have been employed for the isolation of microbial strains capable of degrading azo dyes are summarized in Table 2.

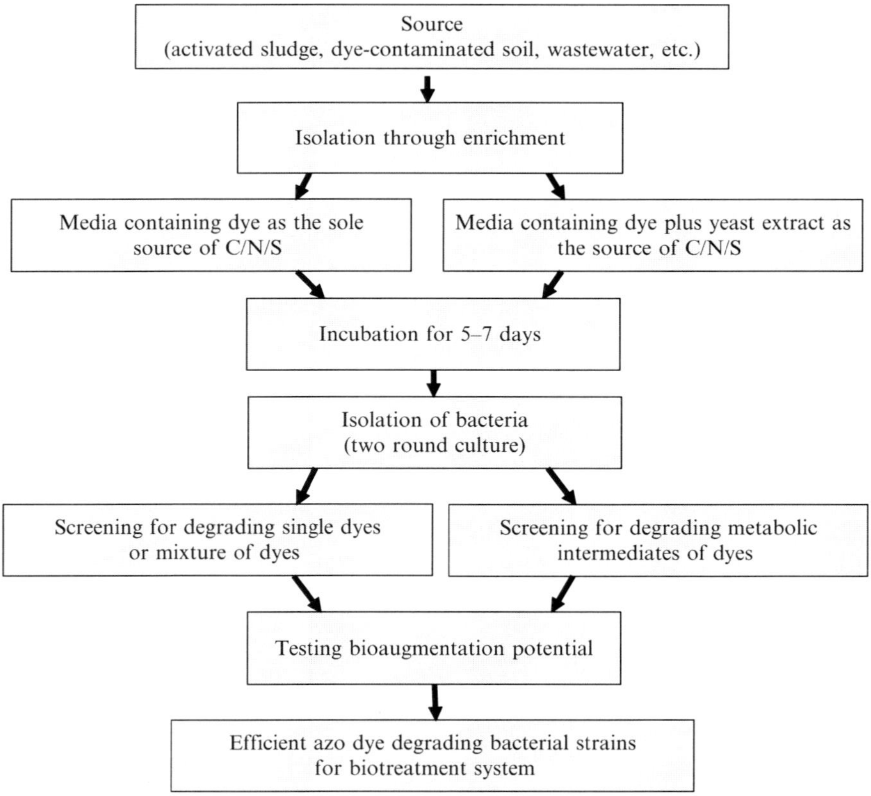

Fig. 2 Key steps for the isolation of efficient azo dye degrading bacterial strains for biotreatment systems

Table 2 Methods for isolation of azo dye-degrading bacteria reported by various authors

Source for isolation	C and/or N source	Identified strains	References
Activated sludge	Reactive Yellow-107, Reactive Black-5, Reactive Red-198, Direct Blue-71, glucose, sodium pyruvate	*Klebsiella* sp. strain VN-31	[39]
Activated sludge	4-Nitroaniline (an intermediate of dye), yeast extract	*Acinotobacter* sp., *Citrobacter freundii*, *Klebsiella oxytoca*	[14]
Activated sludge	Reactive Red-180, glucose	*Citrobacter* sp. CK3	[65]
Activated sludge	Azo dyes, yeast extract	*Bacillus cereus* DC11	[59]
Activated sludge	Azo dyes, glucose	*Staphylococcus arlettae*	[38]
Activated sludge	Beef and yeast extract	*Zoogloea* spp. (from activated sludge) and *Escherichia coli* (GEM)	[54]
Activated sludge of a textile printing wastewater plant	Fast Acid Red GR, Reactive Brilliant Blue, yeast extract	*Shewanella decolorationis* sp. nov.	[85]
Activated sludge, turfgrass soil	Acid Red-88, Reactive Black-5, Direct Red-81, Disperse Orange-3	*Shewanella putrefaciens* AS96, *Aeromonas punctata* AS81, *Bacillus cereus* AS7, *Bacillus thuringiensis* S46, *Pseudomonas nitroreducens* AS77, *Massilia timonae* S81	[40]
Bromoamine acid contaminated soil	Bromoamine acid (an intermediate of anthroquinone dye)	*Sphingomonas herbicidovorans*	[87]
Coastal seawater	Yeast extract, azo dyes	*Vibrio harveyi* TEMS1	[132]
Coastal sediment	Yeast extract, peptone	*Halomonas* sp strain GTW	[133]
Dye contaminated soil	Dye alone or dye with glucose, yeast extract	*Paenibacillus polymyxa*, *Micrococcus luteus*, *Micrococcus* sp.	[76]
Dye contaminated soil and sludge	Acid Orange-7, yeast extract	Consortium consisting of *Aeromonas caviae*, *Proteus mirabilis* and *Rhodococcus globerulus*	[58]
Dye contaminated soil and wastewater	Acid Red-119	*Bacillus thuringiensis*	[63]
Effluent, sludge from textile treatment	Direct Black-38, yeast extract	*Enterococcus gallinarum*	[46, 66]
Not known	Textile dyes	*Pseudomonas* sp.	[79]

(*continued*)

Table 2 (continued)

Source for isolation	C and/or N source	Identified strains	References
Not known	Reactive Black 5, Reactive Yellow 145	*Pseudomonas fluorescens*	[4]
Sludge	Direct Black-38	Not identified	[134]
Sludge	Reactive Black 5, Direct; Brown 2, glucose	*Escherichia coli*, *Pseudomonas* sp.	[135]
Sludge	Glucose or sodium acetate	*Pseudomonas fluorescens*, *Acinetobacter culcoaceticus*	[136]
Sludge, textile effluent treatment plant	Raective Black-5, sodium lactate, yeast extract	*Rhodopseudomonas palustris* W1	[137]
Soil of disposal site of a textile industry	Disperse Blue-79, Acid Orange-10, yeast extract	*Bacillus fusiformis* KMK 5	[60]
Soil of tannery effluent site	Congored	*Bacillus* sp.	[61]
Soil samples from dairy wastewater and from dairy food industries	Methyl Orange	*Lactobacillus casei* TISTR 1500	[75]
Waste disposal site of textile industry	Red BL1/Reactive Red-2, yeast extract, beef extract	*Pseudomonas* sp. SUK1	[82, 138]

As illustrated in Table 2, enrichment culture is the most common method for isolating azo-dye degrading bacteria, using specific dyes individually or in mixtures, where the dyes are provided as the sole source of C or N [40, 79, 87, 139, 140]. Such bacteria cleave azo (–N=N–) bonds reductively and utilize aromatic amines as the source of C and N for their growth and they are specific towards their substrate. On the other hand, other bacterial strains cannot utilize dye as the growth substrate [115], but can be isolated using other organic compounds that are added as a cosubstrate along with the dye to support their growth. The latter method has led to the isolation of many efficient dye-degrading strains [48, 49, 60, 65, 82, 138], but has the disadvantage that the cosubstrate must be added to the wastewater. Depending on the cosubstrate, this can increase the cost of the treatment process. Moreover, addition of cosubstrates to mixed microbial communities containing undefined mixtures of bacterial species from the environment can lead to competition between the inoculant and other bacteria that degrade the cosubstrate.

Activated sludge is usually used as a source of inocula for isolating azo dye degrading bacteria [14, 37–41, 59, 65]. Following isolation of candidate strains, screening under controlled conditions by conducting repeated trials is critical to identify the most effective dye-decolorizing bacterial strains. Similarly, bacterial strains capable of effectively converting/degrading highly toxic intermediates/by-products of dyes can be screened. Ideally a strain or consortium that is able to decolorize azo dyes under anaerobic conditions would also be efficient for further degrading the dye intermediates (aromatic compounds) under aerobic conditions. Finally, strains showing good results under controlled conditions should be tested further for their performance to degrade dyes and their products in a bioaugmented system by co-culturing with bacterial communities from an activated sludge system to determine if the strains are competitive and are able to enhance the dye degradation rates over that which is achieved by a nonaugmented sludge community [14]. Nutritional and other ecological conditions should be optimized for the development of an effective treatment process for the removal of dyes/dye-products from the dye-contaminated wastewaters. Thus, functionality of the selected strains must be defined well before using it as a biotreatment system. This can be achieved by employing a standard set of biochemical and molecular tests in the laboratory. The pure cultures must then be developed into an inoculum that can be stored and transported in a convenient form for delivery to the wastewater treatment facility.

5 System Ecology: Features of Wastewater Treatment Systems

Both biotic and abiotic components of the wastewater treatment systems are crucial considerations in determining whether bioaugmented microbial communities will function effectively for removal of azo dyes from wastewater effluents. Issues related to the success or failure of the treatment systems include adaptation and evolution of the dye-degrading microbial community, activity and interaction with the indigenous microflora, and environmental and nutritional aspects that influence microbial performance in the treatment systems [31, 35, 141–143]. Stability of the azo dye-degrading microbial communities and population dynamics seem to be the most important factors for the stability of the treatment process. However, the processes should not be viewed as a simple function of the microorganisms but as a complex ecosystem composed of a pool of functions contributed by both biotic and abiotic factors. The great diversity observed in such ecosystems has created a challenge for the consistent use of inoculants in assuring complete removal of azo dye contaminants over a wide range of possible environmental conditions. The structure and function of microbial communities often shifts concurrently during its adaptation period in response to fluctuation in the environmental conditions [144–147]. Consequently, performance of the system can be affected dramatically. For this reason, monitoring the composition of the microbial community is vital for the identification of functionally relevant populations [148, 149]. This can be achieved by correlating a specific activity of the process and a typical microbial population to simultaneously examine process performances and microbial population variations.

Fortunately, recent advances in molecular techniques have made it possible for scientists and engineers to monitor dye-degrading communities and their interaction with the other microorganisms during the degradation process (see review: [150]). Before the advent of such techniques, the key microbial species in wastewater treatment plants were either unknown or sometimes inefficient bacteria were considered important for the various processes. Up until the last decade, very few studies employed molecular tools to monitor the degrader communities in activated sludge systems [142, 151–155], whereas more recently the use of such tools has become increasingly common for monitoring microbial compositions and the activity of dye degraders in water treatment systems [42, 147, 150, 156–158]. Ultimately, these techniques should prove useful to identify the links between microbial community composition, function, and process stability. A summary of advanced modern techniques used by scientists to study microbial structures/compositions in the wastewater treatment systems is presented in Table 3.

Prokaryotes that are present in activated sludge or biofilm reactors are responsible for the removal of most of the C and other nutrients or contaminants from wastewater and are the core component of every biological wastewater treatment plant [155]. At the same time, some bacterial species can also be detrimental to the treatment system either by aiding the formation of foam, which affects the settling features of activated sludge, or by out competing or suppressing microbial populations that are responsible for the removal of a particular contaminant. Good settling properties of an activated sludge are crucial for separating treated water from the sludge. Foaming is often caused by excessive growth of filamentous bacteria [150, 169–171]. The flocs containing high amounts of filaments with hydrophobic cell surfaces tend to attach to air bubbles and float on the surface of the sludge basin, from where they are easily dispatched by wind.

6 Bioaugmentation with Azo Dye Degrading Bacteria

Although conventional activated sludge systems are commonly used to treat azo dye containing wastewater [16, 29, 70–72, 172, 173], these treatment systems are inconsistent for removal of recalcitrant azo dyes and are subject to failure due to poor environmental conditions [32, 42, 43, 174]. The bioaugmentation of treatment systems commonly involves the use of mixed cultures of microorganisms (Table 4), and similarly can result in varying treatment efficacy depending on the abilities of the individual strains to compete with indigenous populations that are often well-acclimated to the existing environmental conditions [191]. More recently, individual strains of bacteria have been reported to have exceptional traits and can greatly accelerate dye decolorization rates (Table 5). In addition to azo dye degraders, degradation rates sometimes can also be improved by augmentation with a bacterial sp. with nonessential functions to influence treatment performance [42, 95]. For example, *Escherichia coli* DH5α increases the decolorization efficiency of *P. luteola* even though DH5α is not an active decolorizer of azo dyes among the microbial community. In this case, extracellular metabolites expressed by DH5α stimulated decolorization activity of *P. luteola*. In recent work, genetically engineered microorganisms (GEM) have also received attention for biodegradation studies and been widely applied in bioaugmentation systems [55].

Table 3 Techniques used to study degrading microbial community structures in various wastewater treatment systems

Method/technique	Microbial systems	Parameters studied	References
Fluorescent in situ hybridization (FISH)	Consortium comprised of three bacterial strains capable of degrading several textile dyes including azo dyes in a rotating biological contactor	Rate of dye degradation, pH, BOD, and enzymes involved, and survival of test organisms	[159]
Denaturing gradient gel electrophoresis (DGGE), real-time PCR, and FISH	Sludge from a domestic wastewater treatment plant and *Comamonas testosteroni* I2 *gfp*	Effect of chloro-anilines on activated-sludge reactor functions such as nitrification, carbon removal, and sludge compaction, and sludge community structure, particularly the nitrifying populations	[160]
FISH, terminal restriction-fragment length polymorphism analysis (rRNA-based molecular techniques) and comparative 16S rDNA analysis	Activated sludge systems	The bacterial composition of activated sludge from two laboratory plants with different modes of operation, i.e., anoxic/oxic-[enhanced biological phosphorus removal (EBPR), no nitrification] and Phoredox-system (EBPR, nitrification and denitrification) with particular emphasis on microorganisms responsible for EBPR process	[161]
Ribosomal intergenic spacer analysis (RISA), 16S rRNA gene sequencing and ARDRA	Activated sludge and *Sphingomonas xenophaga* QYY	Degradation of bromoamine acid (an intermediate of anthroquinone dye) and microbial community dynamics	[43, 44]
Replacement series method	*Pseudomonas luteola*, *Escherichia coli* DH5 α	Color removal of dye Reactive Red 22, Study notes competition among the degrader species affecting long term stability	[42]
16S rRNA gene clone library	Wastewater purification bioreactor	Bacterial community structure in the natural circulation bioreactor	[162]

FISH and DGGE	Activated sludge samples from the anoxic and aerobic zones of a laboratory-scale modified Ludzack–Ettinger (MLE) system, *Proteobacteria*, and total Eubacteria	Microbial community structures and genetic diversity of the microbial community present in each of the anoxic and aerobic zones, along with COD and nitrogen mass balances	[163]
Scanning electron microscopy, light microscopy, and confocal laser scanning microscopy together with FISH	Aerobic activated sludge granules (spherical biofilms)	Structure of biofilms, microbial composition of heterogeneous granular biofilms and detection of bacteria, ciliates, and fungi in and on granules	[164]
Randomly amplified polymorphic DNA (RAPD), Enterobacterial repetitive intergenic consensus sequence (ERIC-PCR)	A laboratory-scale anaerobic–anoxic–oxic fixed biofilm system treating coking wastewater	Microbial community structural dynamics, and identification of genomic fragments whose abundance shifts were concomitant to changes in COD removal capacity in a reactor	[165, 166]
RISA and ARDRA	Sludge and *Escherichia coli* JM109 (pGEXAZR)	Removal of dye Acid Red GR by bioaugmented sludge and changes in microbial community in the reactor	[54]
DGGE and clone library analysis	Wastewater treatment plant, Oceanospirillales and Methylococcaceae, Caulobacteraceae, Sphingomonadaceae, and Nitrospirae	Effects of wastewater treatment plant discharge on the ecology of bacterial communities in the sediment of a small, low-gradient stream in South Australia, and the quantification of genes involved in the biogeochemical cycling of carbon and nitrogen	[167]

(*continued*)

Table 3 (continued)

Method/technique	Microbial systems	Parameters studied	References
Single-strand conformation polymorphism (SSCP)	Wastewater bioreactors (including denitrifying and phosphate-removal system, Chinese traditional medicine wastewater treatment system, beer wastewater treatment system, fermentative bio-hydrogen producing system, and sulfate-reduction system)	Microbial community structures, diversity and distribution in different wastewater treatment processes, and relationship between the structures and the status of processes	[157]
SSCP	The reactor inoculated with a microbial consortia obtained from a textile wastewater treatment plant	Color removal and changes in bacterial community profile	[168]
DGGE	Water samples collected from eight sites of three different lake zones and the Global Positioning System	Microbial community composition and relationship between bacterial community structure and environmental factors	[147]

Table 4 Biodegradation of azo dyes and their intermediates by mixed microbial cultures

Dyes/Metabolites	Culture	Comments	References
Acid Orange 10, Acid Red 14 and 18	Anaerobic digester sludge and aeration tank mixed liquor	Decolorization ranging from 65–90% was observed in a two-stage anaerobic–aerobic fixed film fluidized bed activated sludge reactor	[175]
Acid Orange 7	Sludge originally collected from a pulp and paper wastewater treatment plants	Color removal of 96% was achieved in the presence of liposomes that facilitated uptake of dyes by anaerobic biomass, leading to a fast decolorization. Amines such as sulfanilic acid and aniline were mineralized by inocula with high microbiological diversity, even with domestic effluent. Orthanilic and metanilic acids and 1-amino-2-naphtol were persistent under tested conditions	[176]
Acid Orange 7	Granular activated carbon-biofilm configured packed column	With initial 500 mg L^{-1} dye concentration, a complete decolorization was achieved in all runs although the cosubstrates added into the BGAC-packed column system reduced until to zero	[177]
Acid Orange 7	Uncharacterized aerobic biofilm, *Sphinogomonas* sp. 1CX and Gram-negative bacterium strain SAD4I	The dye was completely degraded within 1 h in a rotating drum bioreactor containing the biofilm. The two bacterial strains in co-culture were able to mineralize the dye up to 90%	[178]
Acid Orange 7	Mixed and methanogenic cultures	The culture exhibited 94% color removal. Color removal was faster in mixed cultures than in methanogenic culture. Addition of electron donor stimulated reductive cleavage of azo bond	[179]
Acid Orange 7 and many other dyes	Bacterial consortium TJ-1 consisting of *Aeromonas caviae*, *Proteus mirabilis*, and *Rhodococcus globerulus*	Decolorization of Acid Orange 7 was significantly higher with the consortium as compared to the individual strains. More than 90% decolorization could be achieved even at 200 mg L^{-1} within 16 h. The consortium also decolorized 15 other azo dyes individually as well as a simulated wastewater containing a mixture of all the 16 azo dyes	[58]

(*continued*)

Table 4 (continued)

Dyes/Metabolites	Culture	Comments	References
Acid Red 42, Acid Red 73, Direct Red 80, Disperse Blue 56	Sludge collected from a municipal wastewater treatment plant	Average removal efficiency for acid dyes was between 80 and 90%. The removal efficiency for Direct Red 80 was 81% while of Disperse Blue 56 was not observed	[180]
Acid Red 88, Reactive Black 5, Direct Red 81, Disperse Orange 3	Activated sludge, *Shewanella putrefaciens* AS96, *Aeromonas punctata* AS81	Strains AS81 and AS96 from the activated sludge were able to decolorize all the tested four dyes in liquid medium after bioaugmentation into a live culture of activated sludge. The unamended activated sludge had little capacity to decolorize the dyes with 14% decolorization occurring after 8 h	[40]
Acid Red 88, Reactive Black 5, Direct Red 81, Disperse Orange 3, 4-nitroaniline (an intermediate of dye)	Mixed bacterial culture (*Acinetobacter* sp., *Citrobacter freundii* and *Klebsiella oxytoca*, *Shewanella putrefaciens* AS96) isolated from activated sludge	Under static conditions, 88–100% decolorization of the tested dyes was achieved by mixed bacterial culture after 10 h incubation. The mixed bacterial culture plus *S. putrefaciens* AS96 exhibited complete decolorization in <6 h. Further incubation of the solutions that contained *S. putrefaciens* and the mixed bacterial culture for 48 h under aerobic conditions resulted into complete removal of 4-nitroaniline residues	[14]
Azo dye metabolites	Activated sludge	Under aerobic conditions, two compounds 4,4′-thiodianiline and *p*-kresidine were most easily degraded, followed by 4,4′-diaminodiphenylmethane and 2-naphthylamine. Under anaerobic conditions, stabilities of the amines were totally different and compounds *p*-kresidine, 4,4′-diaminodiphenyl methane, and 2-naphthylamine were degraded while the 4-chloroaniline, 2,4-diaminotoluene, and 2,4-diaminoanisole were partly degraded	[16]
Broamine acid (BAA) and azo dyes	Salt tolerant mixed bacterial culture	The dyes were decolorized only under anaerobic conditions. The BAA could significantly increase the decolorization of one of the test dye by the salt-tolerant bacteria	[133]

(*continued*)

Table 4 (continued)

Dyes/Metabolites	Culture	Comments	References
Direct Black 38	Granulated anaerobic sludge mixed culture	The dye was degraded and decolorized throughout the experimental period of 300 h. The batch anaerobic tests indicated that once reduced environments were established with glucose, decolorization occurs even at high dye concentrations	[181]
Direct Black 38	Mixed microbial culture isolated from an aerobic bioreactor treating textile wastewater	The dye was transformed into benzidine and 4-aminobiphenyl followed by complete biodegradation of these toxic intermediates	[134]
Direct Blue 71	Anaerobic sludge plus *Escherichia coli* JM109	The bioaugmentation improved the removal of the target compound by the sludge. The bioaugmented reactor also demonstrated faster DB 71 decolorization rate than the control one	[55]
Direct Fast Scarlet 4BS	Consortium of a white-rot fungus and *Pseudomonas* 1–10 isolated from wastewater	The microbial consortium showed a significant improvement in dye decolorization rates under either static or shaking culture. The 4BS was mineralized completely	[89]
Direct Red 81	Consortium from contaminated soils in the vicinities of dye-stuff manufacturing units	The consortium exhibited 90% decolorization ability within 35 h	[96]
Dye-containing wastewater and Reactive Red 22	Mixed cultures of *Pseudomonas luteola* and *E. coli* DH5α	Presence of *E. coli* DH5α increased the decolorization efficiency of *P. luteola* even though DH5α was an inefficient decolorizer in this consortium	[42, 95]
Industrial wastewater containing precursors and synthesis products of 15 sulfonated azo dyes	Anaerobic baffled reactor containing mixed sulfate reducing bacteria, and methanogens	In an anaerobic baffled reactor, almost a complete removal of color was observed in the reactor within 100 days of operation	[182]
Mordant Yellow 3, Acid Red 27, Yellow 23 and 21	Mixed bacterial culture	Reduction of dye under anaerobic conditions occurred followed by oxidation of amine metabolites after re-aeration	[183]

(*continued*)

Table 4 (continued)

Dyes/Metabolites	Culture	Comments	References
Orange G, Amido Black 10B, Direct Red 4BS and Congo Red	Four bacterial strains (pseudomonads) isolated from dyeing effluent-contaminated soils	Maximum degradation observed in the treatment system after 24 h for Orange G was 60.9 mg L^{-1}, for Amido Black 10B 571.3 mg L^{-1}, for Direct Red 4BS 112.5 mg L^{-1}, and for Congo Red 134.9 mg L^{-1}	[184]
Orange II and other azo dyes	Biodigester sludge from municipal waste plant augmented with sulfate reducing consortium	The dye was decolorized by 95% within 24 h. Several other dyes including Reactive Black 5 and Reactive Red 120 and mixture of dyes were successfully degraded	T[185]
Procion navy blue, Procion green, Direct blue, and a mixture of azo dyes	Consortium comprised of three bacterial strains	A high efficiency for dye degradation was observed even at high dye concentrations	[159]
Reactive Black 5	Activated sludge batch reactor	Color removal occurred under anaerobic environment, while a slight attenuation was noticed under the aerobic condition	[136]
Reactive Black 5 and Direct Brown 2	Granulated anaerobic sludge mixed culture	Decolorization and substrate removal were achieved under test conditions but ultimate removal of azo dyes and substrate were not observed at high dye concentrations. Aromatic amine and volatile fatty acid accumulation observed proportionally at higher azo dye concentration	[135]
Reactive brilliant red X-3B	Activated sludge	Activated sludge generated a heterogeneous biofilm of several bacterial species. Biomass was increased after ozonation. Combination of ozone oxidation and upflow biological aerated filter technique to treat azo dyes	[8]
Reactive Orange 96	Anaerobic culture of sulfate-reducing bacteria, methanogens, and fermentative bacteria	Sulfate-reducing bacteria removed 95% of the dye in 40 h. Methane producing bacteria did not contribute in dye removal. Fermentative bacteria could remove only 30% of the dye in 90 h	[186]
Reactive Red 3.1	Activated sludge obtained from domestic and industrial effluent treatment plants	Decolorization rates of up to 30 mg L^{-1} h^{-1} were observed in case of activated sludge under anaerobic conditions. In anaerobic packed bed reactor	[15]

(*continued*)

Table 4 (continued)

Dyes/Metabolites	Culture	Comments	References
		followed by aerobic stirred tank reactor 90–93% dye removal occurred after 51 h	
Reactive Violet 5R and several other dyes	Consortium of three bacteria (*Paenibacillus polymyxa*, *Micrococcus luteus*, and *Micrococcus* sp.) isolated from dye-contaminated soil	The concerted metabolic activity of these isolates led to complete decolorization of Reactive Violet 5R (100 mg L^{-1}) within 36 h whereas individual isolates could not show decolorization even on extended incubation. The consortium had the ability to decolorize nine dyes amongst the 10 tested	[76]
Remazol Black B	*Alcaligenes faecalis*, *Commomonas acidovorans*	Microbial consortium immobilized on gravel exhibited over 95% decoloration within 48 h	[187]
Remazol Brilliant Violet 5R and Remazol Black B	Sequencing batch reactor inoculated with sludge collected from activated sludge plant	About 90% color removal was recorded for Remazol Brilliant Violet 5R and 75% color removal was obtained for Remazol Black B in a 24-h cycle with a sludge retention time of 15 days and an aerated reaction phase of 10 h	[188, 189]
Scarlet R	Consortium comprised of *P. vulgaris* and *M. glutamicus*	The consortium completely decolorized the dye in 3 h, the time was much shorter than the pure cultures	[77]
Simulated textile wastewater containing Procion Red H-E7B	Inclined Tubular Digester granules from pulp plant (upflow anaerobic sludge blanket)	A 78% color removal by anaerobic treatment was observed. Upflow anaerobic sludge blanket gave better color removal than inclined tubular digester	[190]
Textile wastewater containing Reactive Red 120	Activated sludge and fermented sludge of municipal wastewater	Over 90% decolorization was obtained on anaerobic phase of the bioreactor	[17]

Bioaugmentation of activated sludge systems with efficient bacterial strains can be used to target both the parent compounds and their degradation products such as aromatic hydrocarbons [29, 30, 32, 55]. To be effective, such strains should meet at three criteria [33]: (1) they must be catabolically active, (2) they must be competitive to sustain a high population density after being introduced into the system, and (3) they should be compatible with indigenous microbial communities and should not affect the indigenous microbial communities adversely. Thus candidate bacteria should be carefully evaluated with respect to each criterion. Several studies have identified potential candidates for use in bioaugmentation [40, 42–44, 54, 55].

Table 5 Rate of azo dye decolorization by different microbial strains

Dye	Strain	Dye concentration	Decolorization rate (mg dye h^{-1})	References
Acid Red 119	*Bacillus thuringiensis*	300 mg L^{-1}	218	[63]
Acid Red 88, Direct Red 81, Reactive Black 5, Disperse Orange 3	*Shewanella putrefaciens* AS96	100 mg L^{-1}	22.1–25.0	[41]
Direct Black 38	*Enterococcus gallinarum*	491 mg L^{-1}	12.8	[64]
Direct Fast Scarlet 4BS	Bacterial and fungal consortium	1,000 mg L^{-1}	81.2	[89]
Direct Red 28	*Bacillus velezensis*	25 mg L^{-1}	2.5	[66]
Direct Red 81	Bacterial consortium	100 mg L^{-1}	2.5	[96]
Methyl Orange, Methyl Red	*Lactobacillus casei* TISTR 1500	0.23 mmol L^{-1}	6.1–31.0	[75]
Reactive Red 22	*Escherichia coli* NO3	200 mg L^{-1}	17.0	[67]
Scarlet R	Consortium GR comprised of *Proteus vulgaris* and *Micrococcus glutamicus*	50 mg L^{-1}	16.7	[77]

Recently, we demonstrated that the strains of genus *Shewanella* could potentially be useful for the treatment of azo dyes. One such isolate, *Shewanella putrefaciens* strain AS96, which was purified from an activated sludge was able to decolorize four structurally different azo dyes (Acid Red-88, Reactive Black-5, Direct Red-88, and Disperse Orange-3) in a liquid medium and maintained a high catabolic rate when introduced into a mixed microbial community from activated sludge [14, 40]. The rate of dye decolorization was nearly identical for the pure culture as for the bioaugmented sludge (Fig. 3). Similarly, bacterial strains belonging to genus *Sphingomonas* have been shown to degrade azo dyes [52, 53]. One strain identified as *S. xenophaga* QYY was used to degrade an intermediate of anthroquinone dye, bromoamine acid (BAA) [43, 44] and was suggested as a good candidate for bioaugmentation to remove BAA in laboratory sequencing batch reactors.

Nutritional and environmental factors such as C and energy source, redox mediators, salinity, temperature, pH, and oxygen supply affect the biomass and degradation activity of azo dye degrading microorganisms [10, 14, 40, 41, 69, 76, 84, 109, 133, 192–194]. Provision of optimal conditions can therefore enhance the effectiveness and success of the azo dye bioaugmented treatment systems. When bacteria are introduced into a complex microbial community, the nutritional conditions are altered by competition with the indigenous microflora, and monitoring of biomass and population size over time is complicated since specific methods are

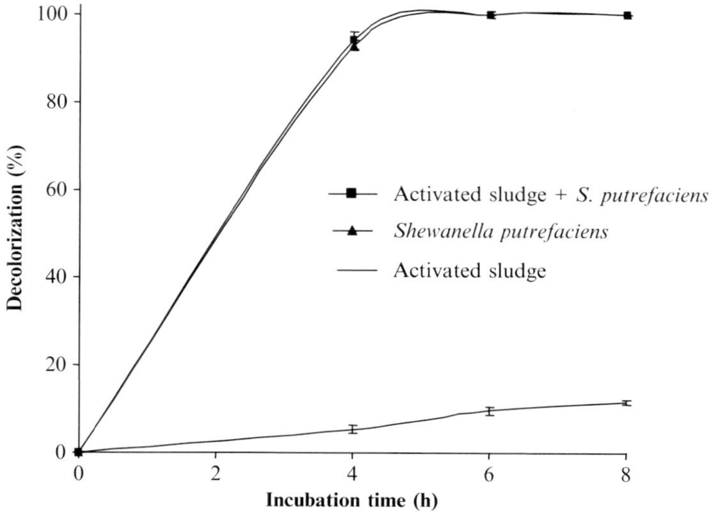

Fig. 3 Decolorization of Reactive Black-5 by activated sludge after augmentation with Shewanella putrefaciens AS96
Source: [40, 41]

needed to monitor the introduced strains. This can be achieved by using molecular techniques as described in Table 3 to assess the persistence and activity of the augmented bacteria in the presence of the indigenous population.

One of the main considerations in degrading azo dyes is the effect of oxygen at different stages in the process. Normally decolorization is achieved under low oxygen conditions, which results in the production of potentially toxic metabolites [13, 24, 176, 195, 196]. The later products can then be degraded by switching to high oxygen conditions. Thus, biological processes for azo dye degradation can require sequential anaerobic–aerobic conditions. This can be achieved either in a single reactor by careful aeration control for different periods or in two separate reactors [51, 197]. Although pilot-scale and fullscale implementation of anaerobic–aerobic biological treatments are limited, very promising results have been reported by the scientists using this system for the removal of dyes and their toxic products [14, 84, 86, 198, 199].

Another critical issue is the presence of a high concentrations of salt in dye-contaminated textile effluents that may affect azo dye degrading microorganisms by causing plasmolysis of the cells, by lowering metabolic activity, or by conformational changes in the degradation enzymes. Among the hundreds of studies on biodegradation of azo dyes, relatively few studies have employed high salt conditions, especially at $\geq$10% salt concentration [15, 41, 133, 200–205]. Therefore, biological treatment systems require exploitation of microbial species that can thrive and degrade azo dyes at high salt concentrations.

7 Practical Considerations and Future Outlook

Several studies have demonstrated unequivocally that bioaugmentation with selected bacteria can be used to facilitate the degradation of azo dye compounds in wastewater (Table 1). However, several practical considerations need to be taken into account to develop inocula that can be used at the field scale. These include (1) effective cell densities, (2) ease in production of the inoculum, (3) inoculum carrier or delivery system, (4) shelf life, and (5) survival and adaptability of active degraders in the treatment system. The inoculum production system should maintain a metabolically and physiologically competent state to obtain desired benefits. Stringent quality assurance at various steps of production and packaging is required for the production of high quality inoculants. It is imperative that the formulation should be cost-effective and stable during production, distribution, storage, and transportation. Moreover, the formulation should be easy to handle and apply so that it is delivered to the target in the most appropriate manner and form.

Maintaining high levels of specific inoculants after bioaugmentation in the treatment systems is a great challenge since the inoculants compete with indigenous microbiota for growth factors, are subject to starvation, predation by protozoa, and washout at high flow rates. To prevent washout of cells and maintain a high concentration of cells in the bioreactor, various systems including submerged-membrane bioreactor and immobilized bioreactor systems are often employed [206]. The treatment of wastewater in packed bed bioreactors using immobilized cells is receiving more attention with the application of different immobilization methods and a variety of carriers [70–72, 177, 207–212]. Compared to conventional free cell systems, the bioreactors with immobilized cells have shown better results in terms of reactor productivity and ability to withstand extreme environments [213, 214]. In the immobilized bioreactor systems, various support media such as granular activated carbon, polyurethane foam, and ceramics are used to enhance the performance of immobilized cells on a long term basis. The application of such carriers in bioaugmentation systems is now viewed as a promising approach for the retention of sufficient biomass and for the prevention of washout of cells under high flow rates [208]. Among the various carriers, ceramic immobilized systems are the most durable and have been well suited for anaerobic treatment of wastewater [215].

Another strategy that has not been employed yet would be to use inocula as vectors to introduce catabolic pathways for azo dye degradation into the indigenous community in the waste water stream. Degradation pathways are frequently carried on plasmids and transposons that integrate into the chromosome as mobile genetic elements (MGEs). The MGEs, which can even mediate their own horizontal gene transfer, can play a major role in bacterial adaptation [216, 217]. Various mechanisms of horizontal gene transfer have been documented [216, 218–221]. Conceivably, MGEs may move freely within the bacterial community, although eventual expression of the genes may depend on compatibility of the promoters and integration into the regulatory systems in different bacterial species. Both biotic and abiotic factors including competition between the organisms, predation, nutrients, temperature, pH, oxygen, etc. also will influence horizontal gene transfer [217, 222, 223]. Plasmids with broad host range permit interspecies genetic exchange and may, therefore, be a major factor for the adaptation of microbial communities. In this case, marker genes or genetic sequence information on the components of the gene pool could be useful to identify the distribution of the MGEs within different species in mixed microbial cultures.

The performance of a biotreatment system ultimately depends on optimization of the activity of microbes and the ability to control the process parameters of the treatment system [157]. In this respect, the ability to monitor gene copy numbers and gene expression is highly useful for real time optimization of the efficiency of a biotreatment system. Advanced molecular techniques as well as low cost methods (e.g., antibody detection of enzymes based on color reaction strips; fluorescence i.e., GFP marked organisms with UV light detection) can also be applied to monitor the microbial community structure, persistence of the added bacteria, and their interactions with indigenous populations.

References

1. Alinsafi A, Evenou F, Abdulkarim EM et al (2007) Treatment of textile industry wastewater by supported photocatalysis. Dyes Pigm 74:439–445
2. Arslan-Alaton I (2007) Degradation of a commercial textile biocide with advanced oxidation processes and ozone. J Env Manage 82:145–154
3. Behnajady MA, Modirshahla N, Shokri M (2004) Photodestruction of Acid Orange 7 (AO7) in aqueous solutions by UV/H2O2: influence of operational parameters. Chemosphere 55:129–134
4. Brosillon S, Djelal H, Merienne N, Amrane A (2008) Innovative integrated process for the treatment of azo dyes: coupling of photocatalysis and biological treatment. Desalination 222:331–339
5. El-Gohary FA, Badawy MI, El-Khateeb MA, El-Kalliny AS (2009) Integrated treatment of olive mill wastewater (OMW) by the combination of Fenton's reaction and anaerobic treatment. J Hazard Mat 162:1536–1541
6. Golab V, Vinder A, Simonic M (2005) Efficiency of the coagulation/flocculation method for the treatment of dye bath effluent. Dyes Pigm 67:93–97
7. Hao JJ, Song FQ, Huang F, Yang CL, Zhang ZJ, Zheng Y, Tian XJ (2007) Production of laccase by a newly deuteromycete fungus *Pestalotiopsis* sp. and its decolorization of azo dye. J Industr Microbiol Biotechnol 34:233–240
8. Lu X, Yang B, Chen J, Sun R (2009) Treatment of wastewater containing azo dye reactive brilliant red X-3B using sequential ozonation and upflow biological aerated filter process. J Hazard Mat 161:241–245
9. Moustafa ME (2005) Synthesis and structural and biological activity studies on some lanthanide chelates with O- and N-containing ligands. Spectr Lett 38:23–34
10. Pearce CI, Lloyd JR, Guthrie JT (2003) The removal of colour from textile wastewater using whole bacterial cells: a review. Dyes Pigm 58:179–196
11. Saxe JP, Lubenow BL, Chiu PC, Huang CP, Cha DK (2006) Enhanced biodegradation of azo dyes using an integrated elemental iron-activated sludge system effects of physical-chemical parameters. Wat Env Res 78:26–30

12. Wang A, Qu J, Liu H, Ge J (2004) Degradation of azo dye Acid Red 14 in aqueous solution by electrokinetic and electooxidation process. Chemosphere 55:1189–1196
13. Dubrow SF, Boardman GD, Michelsen DL (1996) Chemical pretreatment and aerobic-anaerobic degradation of textile dye wastewater. In: Reife A, Freeman HS (eds) Environmental chemistry of dyes and pigments. Wiley, New York
14. Khalid A, Arshad M, Crowley DE (2009) Biodegradation potential of pure and mixed bacterial cultures for removal of 4-nitroanaline from textile dye wastewater. Wat Res 43:1110–1116
15. Bromley-Challenor KCA, Knapp JS, Zhang Z et al (2000) Decolorization of an azo dye by unacclimated activated sludge under anaerobic conditions. Wat Res 34:4410–4418
16. Ekici P, Leupold G, Parlar H (2001) Degradability of selected azo dye metabolites in activated sludge systems. Chemosphere 44:721–728
17. Paździor K, Klepacz-Smółka A, Ledakowicz S, Sójka-Ledakowicz J, Mrozińska Z, Żyłła R (2009) Integration of nanofiltration and biological degradation of textile wastewater containing azo dye. Chemosphere 75:250–255
18. Worch E, Grischek T, Bomick H, Eppinger P (2002) Laboratory tests for simulating attenuation processes of aromatic amines in riverbank filtration. J Hydrol 266: 259–268
19. Cartwright RA (1983) Historical and modern epidemiological studies on populations exposed to N-substituted aryl compounds. Env Health Persp 49:13–19
20. Chung KT, Cerniglia CE (1992) Mutagenicity of azo dyes: structure–activity relationships. Mut Res 277:201–220
21. Makinen PM, Theno TJ, Ferguson JF, Ongerth JE, Puhakka JA (1993) Chlorophenol toxicity removal and monitoring in aerobic treatment: recovery from process upsets. Env Sci Technol 27:1434–1439
22. Miller JA, Miller EC (1983) Some historical aspects of N-aryl carcinogens and their metabolic activation. Env Health Persp 49:3–12
23. Ozturk A, Abdullah MI (2001) Toxicological effect of indole and its azo dye derivatives on some microorganisms under aerobic conditions. Sci Total Env 358:137–142
24. Pinheiro HM, Touraud E, Thomas O (2004) Aromatic amines from azo dye reduction: status review with emphasis on direct UV spectrophotometric detection in textile industry wastewaters. Dyes Pigm 61:121–139
25. Saupe A (1999) High rate biodegradation of 3- and 4-nitroaniline. Chemosphere 39:2325–2346
26. Weisburger JH (2002) Comments on the history and importance of aromatic and heterocyclic amines in public health. Mut Res 506–507:9–20
27. Carliell CM, Barclay SJ, Naidoo N et al (1994) Anaerobic decolorisation of reactive dyes in conventional sewage treatment processes. Wat SA 20:341–344
28. Manu B, Chauhari S (2003) Decolorization of indigo and azo dyes in semicontinuous reactors with long hydraulic retention time. Process Biochem 38:1213–1221
29. Boon N, Goris J, de Vos P, Verstraete W, Top EM (2000) Bioaugmentation of activated sludge by an indigenous 3-chloroaniline-degrading *Comamonas testosterone* strain 12gfp. Appl Env Microbiol 66:2906–2913
30. McClure NC, Fry JC, Weightman AJ (1991) Survival and catabolic activity of natural and genetically engineered bacteria in laboratory-scale activated sludge unit. Appl Env Microbiol 57:366–373
31. Rittman BE, Whiteman R (1994) Bioaugmentation: a coming of age. Biotechnol 1:12–16
32. van Limbergen HV, Top EM, Verstrate W (1998) Bioaugmentation in activated sludge: current features and future perspectives. Appl Microbiol Biotechnol 50:16–23
33. Yu ZT, Mohn WW (2001) Bacterial diversity and community structure in an aerated lagoon revealed by ribosomal intergenic spacer analyses and 16S ribosomal DNA sequencing. Appl Environ Microbiol 67:1565–1574
34. Limbergen HV, Top EM, Verstrate W (1998) Bioaugmentation in activated sludge: current features and future perspectives. Appl Microbiol Biotechnol 50:16–23

35. Dabert P, Delgenes JP, Moletta R, Godon JJ (2002) Contribution of molecular microbiology to the study in water pollution removal of microbial community dynamics. Rev Env Sci Biotechnol 1:39–49
36. Lucas MS, Amaral C, Sampaio A, Peres JA, Dias AA (2006) Biodegradation of the diazo dye Reactive Black 5 by a wild isolate of *Candida oleophila*. Enzyme Microbial Technol 39:51–55
37. Chunli Z, Jiti Z, Jing W, Jing W, Baocheng Q (2008) Isolation and characterization of a nitrobenzene degrading yeast strain from activated sludge. J Hazard Mat 160:194–199
38. Elisangela F, Andrea Z, Fabio DG et al (2009) Biodegradation of textile azo dyes by a facultative *Staphylococcus arlettae* strain VN-11 using a sequential microaerophilic/aerobic process. Int Biodeter Biodegr 63:280–288
39. Franciscon E, Zille A, Garboggini FF et al (2009) Microaerophilic-aerobic sequential decolourization/biodegradation of textile azo dyes by a facultative *Klebsiella* sp. strain VN-31. Process Biochem 44:446–452
40. Khalid A, Arshad M, Crowley DE (2008) Accelerated decolorization of structurally different azo dyes by newly isolated bacterial strains. Appl Microbiol Biotechnol 78:361–369
41. Khalid A, Arshad M, Crowley DE (2008) Decolorization of azo dyes by *Shewanella* sp. under saline conditions. Appl Microbiol Biotechnol 79:1053–1059
42. Chen BY, Chen SY, Lin MY, Chang JS (2006) Exploring bioaugmentation strategies for azo-dye decolorization using a mixed consortium of *Pseudomonas luteola* and *Escherichia coli*. Process Biochem 41:1574–1581
43. Qu Y, Zhou J, Wang J (2006) Bioaugmentation of bromoamine acid degradation with Sphingomonas xenophaga QYY and DNA fingerprint analysis of augmented systems. Biodegr 17:83–91
44. Qu Y, Zhou J, Wang J, Xiang F, Xing L (2005) Microbial community dynamics in bioaugmented sequencing batch reactors for bromoamine acid removal. Microbiol Lett 246:143–149
45. Apohan E, Yeslada O (2005) Role of white rot fungus *Funalia trogii* in detoxification of textile dyes. J Basic Microbiol 45:99–105
46. Bafana A, Chakrabarti T, Muthal P, Kanade G (2009) Detoxification of benzidine-based azo dye by E. gallinarum: Time-course study. Ecotoxicol Env Safety 72:960–964
47. Jadhav JP, Govindwar SP (2006) Biotransformation of malachite green by *Saccharomyces cerevisiae* MTCC 463. Yeast 23:315–323
48. Khehra MS, Saini HS, Sharma DK et al (2005) Decolorization of various azo dyes by bacterial consortia. Dyes Pigm 67:55–61
49. Khehra MS, Saini HS, Sharma DK, Chadha BS, Chimni SS (2005) Comparative studies on potential of consortium and constituent pure bacterial isolates to decolorize azo dyes. Wat Res 39:5135–5141
50. Kornillowicz-Kowalska TK, Wrzosek M, Ginalska G, Iglik H, Bancerz R (2006) Identification and application of a new fungal strain *Bjerkandera audusta* R59 in decolorization of daunomycin wastes. Enz Microbial Technol 38:583–590
51. Pandey A, Singh P, Iyengar L (2007) Bacterial decolorization and degradation of azo dyes. Int Biodeter Biodegr 59:73–84
52. Busse HJ, Kampfer P, Denner EBM (1999) Chemotaxonomic characterization of *Sphingomonas*. J Industr Microbiol Biotechnol 23:242–251
53. David CW, Susan DS, David BR (1996) The genus *Sphingomonas*: physiology and ecology. Curr Opinion Biotechnol 7:301–306
54. Jin R, Zhou J, Zhang A, Wang J (2008) Bioaugmentation of the decolorization rate of acid red GR by genetically engineered microorganism *Escherichia coli* JM109 (pGEX-AZR). World J Microbiol Biotechnol 24:23–29
55. Jin R, Yang H, Zhang A, Wang J, Liu G (2009) Bioaugmentation on decolorization of C.I. Direct Blue 71 by using genetically engineered strain *Escherichia coli* JM109 (pGEX-AZR). J Hazard Mat 163:1123–1128

56. Wagner M, Loy A, Nogueria R, Purkhold U et al (2002) Microbial community composition and function in wastewater treatment plants. Antonie van Leeuwenhoek Int J Gen Mol Microbiol 81:665–680
57. Ghodake G, Jadhav S, Dawkar V, Govindwar S (2009) Biodegradation of diazo dye Direct brown MR by *Acinetobacter calcoaceticus* NCIM 2890. Int Biodeter Biodegr. doi:10.1016/j.ibiod.2008.12.002
58. Joshi T, Iyengar L, Singh K, Garg S (2008) Isolation, identification and application of novel bacterial consortium TJ-1 for the decolourization of structurally different azo dyes. Bioresour Technol 99:7115–7121
59. Deng D, Guo J, Zeng G, Sun G (2008) Decolorization of anthraquinone, triphenylmethane and azo dyes by a new isolated *Bacillus cereus* strain DC11. Int Biodeter Biodegr 62:263–269
60. Kolekar YM, Pawar SP, Gawai KR, Lokhande PD, Shouche YS, Kodam KM (2008) Decolorization and degradation of Disperse Blue 79 and Acid Orange 10, by *Bacillus fusiformis* KMK5 isolated from the textile dye contaminated soil. Bioresour Technol 99:8999–9003
61. Gopinath KP, Meera Sahib HA, Muthukumar K, Velan M (2009) Improved biodegradation of congored by using *Bacillus* sp. Bioresour Technol 100:670–675
62. Mabrouk MEM, Yousef HY (2008) Decolorization of Fast Red by *Bacillus subtilis* HM. J Appl Sci Res 4:262–268
63. Dave SR, Dave RH (2009) Isolation and characterization of *Bacillus thuringiensis* for Acid red 119 dye decolourization. Bioresour Technol 100:249–253
64. Bafana A, Chakrabarti T, Devi SS (2008) Azoreductase and dye detoxification activities of *Bacillus velezensis* strain AB. Env Biotechnol 77:1139–1144
65. Wang H, Su JQ, Zheng XW (2009) Bacterial decolourization and degradation of the reactive dye reactive red 180 by Citrobacter sp. CK3. Int Biodeter Biodegr 63:395–399
66. Bafana A, Krishnamurthi K, Devi SS, Chakarbrati T (2008) Bilogical decolourization of C.I. direct black 38 by *Enterococcus gallinarum*. J Hazard Mat 157:187–193
67. Chang JS, Kuo TS, Chao YP, Ho JY, Lin PJ (2000) Azo dye decolorization with a mutant *Escherichia coli* strain. Biotechnol Lett 22:807–812
68. Isik M, Sponza DT (2003) Effect of oxygen on decolorization of azo dyes by *Escherichia coli* and *Pseudomonas* sp. and fate of aromatic amines. Process Biochem 38:1183–1192
69. Liu G, Zhou J, Wang J (2009) Acceleration of azo dye decolorization by using quinone reductase activity of azoreductase and quinone redox mediator. Bioresour Technol 100:2791–2795
70. Guo J, Zhou J, Wang D et al (2008) The new incorporation bio-treatment technology of bromoamine acid and azo dyes wastewater under high salt conditions. Biodegr 19:93–98
71. Guo J, Zhou J, Wang D, Tian C, Wang Ping M, Uddin S (2008) A novel moderately halophilic bacterium for decolorizing azo dye under high salt condition. Biodegr 19:15–19
72. Guo J, Fang MA, Jiang K, Cui D (2008) Bioaugmentation combined with biofilim process in the treatment of petrochemical wastewater at low temperatures. J Wat Resour Protect 1:1–65
73. Asad S, Amoozegar MA, Pourbabaee AA et al (2007) Decolorization of textile azo dyes by newly isolated halophilic and halotolerant bacteria. Bioresour Technol 98:2082–2088
74. Vijaykumar MH, Vaishampayan PA, Shouche YS, Karegoudar TB (2007) Decolourization of naphthalene-containing sulfonated azo dyes by *Kerstersia* sp. strain VKY1. Enz Microbial Technol 40:204–211
75. Seesuriyachan P, Takenaka S, Kuntiya A (2007) Metabolism of azo dyes by *Lactobacillus casei* TISTR 1500 and effects of various factors on decolorization. Wat Res 41:985–992
76. Moosvi S, Kher X, Madamwar D (2007) Isolation, characterization and decolorization of textile dyes by a mixed bacterial consortium JW-2. Dyes Pigm 74:723–729
77. Saratale RG, Saratale GD, Kalyani DC, Chang JS, Govindwar SP (2009) Enhanced decolorization of textile azo dye Scarlet R by using developed microbial consortium-GR. Bioresour Technol 100:2493–2500

78. Hu TL (2001) Kinetics of azoreductase and assessment of toxicity of metabolic products from azo dyes by *Pseudomonas luteola*. Wat Sci Technol 43:261–269
79. Silveira E, Marques PP, Silva SS (2009) Selection of Pseudomonas for industrial textile dyes decolourization. Int Biodeter Biodegr 63:230–235
80. Kalme S, Jadhav S, Jadhav M, Govindwar S (2009) Textile dye degrading laccase from *Pseudomonas desmolyticum* NCIM 2112. Enz Microbial Technol 44:65–71
81. Ben-Mansour H, Mosrati R, Corroler D et al (2009) In vitro mutagenicity of Acid Violet 7 and its degradation products by *Pseudomonas putida* mt-2: correlation with chemical structures. Env Toxicol Phamacol 27:231–236
82. Kalyani DC, Telke AA, Dhanve RS, Jadhav JP (2009) Ecofriendly biodegradation and detoxification of Reactive Red 2 textile dye by newly isolated *Pseudomonas* sp. SUK1. J of Hazard Mat 163:735–742
83. Wang X, Cheng X, Sun D, Hong Q (2008) Biodecolorization and partial mineralization of Reactive black 5 by a strain of *Rhodopseudomonas palustris*. J Env Sci 20:1218–1225
84. Xu M, Guo J, Sun G (2007) Biodegradation of textile azo dye by *Shewanella decolorationis* S12 under microaerophilic conditions. Appl Microbiol Biotechnol 76:719–726
85. Xu M, Guo J, Cen Y et al (2005) *Shewanella decolorationis* sp. nov., a dye-decolorizing bacterium isolated from activated sludge of a wastewater treatment plant. Int J Syst Evol Microbiol 55:363–368
86. Pearce CI, Christie R, Boothman C et al (2006) Reactive azo dye reduction by *Shewanella* strain J18 143. Biotechnol Bioengin 95:692–703
87. Fan L, Zhu S, Liu D, Ni J (2009) Decolorization of 1-amino-4-bromoanthraquinone-2-sulfonic acid by a newly isolated strain of *Sphingomonas herbicidovorans*. Int Biodeter Biodegr 63:88–92
88. Chang JS, Chen BY, Lin YC (2004) Stimulation of bacterial decolorization of an azo dye by extracellular metabolites from *Escherichia coli* strain NO3. Bioresour Technol 91:243–248
89. He F, Hu W, Li Y (2004) Biodegradation mechanisms and kinetics of azo dye 4BS by a microbial consortium. Chemosphere 57:293–301
90. Nigam P, Banat IM, Singh D, Marchant R (1996) Microbial process for the decolorization of textile effluent containing azo, diazo and reactive dyes. Process Biochem 31:435–442
91. Hong Y, Xu M, Guo J et al (2007) Respiration and growth of *Shewanella decolorations* S12 with an azo compound as the sole electron acceptor. Appl Env Microbiol 73:64–72
92. Chang JS, Lin YC (2001) Decolorization kinetics of recombinant *E. coli* strain harboring azo dye decolorization determinants for *Rhodococcus* sp. Biotechnol Lett 23:631–636
93. Chang JS, Chou C, Lin Y, Ho J, Hu TL (2001) Kinetic characteristics of bacterial azo-dye decolorization by *Pseudomonas luteola*. Wat Res 35:2041–2850
94. Chen KC, Huang WT, Wu YJ, Houng JY (1999) Microbial decolorization of azo dyes by *Proteus mirabilis*. J Industr Microbiol Biotechnol 23:686–690
95. Chen KC, Wu JY, Liou DJ, Hwang SJ (2003) Decolorization of textile dyes by newly isolated bacterial strains. J Biotechnol 101:57–68
96. Junnarkar N, Murty DS, Bhatt NS, Madamwar D (2006) Decolorization of diazo dye Direct Red 81 by a novel bacterial consortium. World J Microbiol Biotechnol 22:163–168
97. Kalme SD, Parshetti GK, Jadhav SU, Govindwar SP (2007) Biodegradation of benzidine based dye Direct Blue-6 by *Pseudomonas desmolyticum* NCIM 2112. Bioresour Technol 98:1405–1410
98. Kapdan IK, Kargi F, McMullan G, Marchant R (2000) Decolorization of textile dye stuffs by a mixed bacterial consortium. Biotechnol Lett 22:1179–1181
99. Moosvi S, Keharia H, Madamawar D (2005) Decolorization of textile dye Reactive Violet 5 by a newly isolated bacterial consortium RVM11.1. World J Microbiol Biotechnol 21:667–672
100. Padmavathy S, Sandhya S, Swaminathan K (2003) Comparison of decolorization of reactive azo dyes by microorganisms isolated from various sources. J Env Sci 15:628–633

101. Yu J, Wang X, Yue PL (2001) Optimal decolorization and kinetic modeling of synthetic dyes by *Pseudomonas* strains. Wat Res 35:3579–3586
102. Chen H (2006) Recent advances in azo dye degrading enzyme research. Curr Protein Pept Sci 7:101–111
103. Blümel S, Knackmuss HJ, Stolz A (2002) Molecular cloning and characterization of the gene coding for the aerobic azoreductase from *Xenophilus azovorans* KF46F. Appl Env Microbiol 68:3948–3955
104. Russ R, Rau J, Stolz A (2000) The function of cytoplasmic flavin reductases in the reduction of azo dyes by bacteria. Appl Env Microbiol 66:1429–1434
105. Keck A, Klein J, Kudlich M et al (1997) Reduction of azo dyes by redox mediators originating in the naphthalenesulfonic acid degradation pathway of *Sphingomonas* sp. strain BN6. Appl Environ Microbiol 63:3684–3690
106. Kudlich M, Keck A, Klein J, Stolz A (1997) Localization of the enzyme system involved in anaerobic reduction of azo dyes by Sphingomonas sp. strain BN6 and effect of artificial redox mediators on the rate of azo dye reduction. Appl Env Microbiol 63:3691–3694
107. Robinson T, McMullan G, Marchant R, Nigam P (2001) Remediation of dyes in textile effluent: a critical review on current treatment technologies with a proposed alternative. Biresour Technol 77:247–255
108. Rau J, Knackmuss HJ, Stolz A (2002) Effects of different quinoid redox mediators on the anaerobic reduction of azo dyes by bacteria. Env Sci Technol 36:1497–1504
109. Brigé A, Motte B, Borloo J, Buysschaert G, Devreese B, Jozef J, Beeumen V (2008) Bacterial decolorization of textile dyes is an extracellular process requiring a multicomponent electron transfer pathway. Microbial Biotechnol 1:40–52
110. Kudlich M, Hetheridge MJ, Knackmuss HJ, Stolz A (1999) Autoxidation reactions of different aromatic o-aminohydroxynaphthalenes that are formed during the anaerobic reduction of sulfonated azo dyes. Env Sci Technol 33:869–901
111. Rafii F, Cerniglia CE (1993) Comparison of the azoreductase and nitroreductase from *Clostridium perfringens*. Appl Env Microbiol 59:1731–1734
112. Rafii F, Cerniglia CE (1995) Reduction of azo dyes and nitroaromatic compounds by bacterial enzymes from the human intestinal tract. Env Health Persp 103:17–19
113. Rafii F, Franklin W, Cerniglia CE (1990) Azoreductase activity of anaerobic bacteria isolated from human intestinal microflora. Appl Env Microbiol 56:2146–2151
114. Rafii F, Smith DB, Benson RW, Cerniglia CE (1992) Immunological homology among azoreductases from *Clostridium* and Eubacterium strains isolated from human intestinal microflora. J Basic Microbiol 32:99–105
115. Stolz A (2001) Basic and applied aspects in the microbial degradation of azo dyes. Appl Microbiol Biotechnol 56:69–80
116. Fewson CA (1988) Biodegradation of xenobiotics and other persistent compounds: the causes of recalcitrance. Trends Biotechnol 6:148–153
117. Ganesh R, Boardman GD, Michelson D (1994) Fate of azo dyes in sludges. Wat Res 28:1367–1376
118. Pagga U, Brown D (1986) The degradation of dyestuffs: part II. behaviour of dyestuffs in aerobic biodegradation tests. Chemosphere 15:479–491
119. Pagga U, Taeger K (1994) Development of a method for adsorption of dyestuffs on activated sludge. Wat Res 28:1051–1057
120. Shaul GM, Holdsworth TJ, Dempsey CR, Dostal KA (1991) Fate of water-soluble azo dyes in the activated sludge process. Chemosphere 22:107–119
121. Ghosh DK, Ghosh S, Sadhukhan P, Mandal A, Chaudhuri J (1993) Purification of two azoreductases from *Escherichia coli* K12. Ind J Exp Biol 31:951–954
122. Ghosh DK, Mandal A, Chaudhuri J (1992) Purification and partial characterization of two azoreductases from *Shigella dysenteriae* type 1. FEMS Microbiol Lett 98:229–234
123. Suzuki T, Timofei S, Kurunczi L et al (2001) Correlation of aerobic biodegradability of sulfonated azo dyes with the chemical structure. Chemosphere 45:1–9

124. Zimmermann T, Gasser F, Kulla H, Leisinger T (1984) Comparison of two bacterial azoreductases acquired during adaptation to growth on azo dyes. Arch Microbiol 138:37–43
125. Zimmermann T, Kulla H, Leisinger T (1982) Properties of purified orange II-azoreductase, the en zyme initiating azo dye degradation by Pseudomonas KF46. Eur J Biochem 129:197–203
126. Hitz HR, Huber W, Reed RH (1978) The absorption of dyes on activated sludge. J Soc Dyers Colorists 94:71–76
127. Chen H, Wang RF, Cerriglia CE (2004) Molecular cloning, overexpression, purification, and characterization of an aerobic FMN-dependent azoreductase from *Enterococcus faecalis*. Protein Exp Purif 34:302–310
128. Walker R, Ryan AJ (1971) Some molecular parameters influencing rate of reduction of azo compounds by intestinal microflora. Xenobiotica 4–5:483–486
129. Beydilli MI, Pavlostathis SG, Tincher WC (2000) Biological decolorization of the azo dye Reactive Red 2 under various oxidation-reduction conditions. Wat Env Res 72:698–705
130. Kulla HG (1981) Biodegradation of synthetic organic colorants. In: Leisinger T, Hutter R, Cook AM, Nuesch J (eds) Microbial degradation of xenobiotics and recalcitrant compounds: FEMS Symposium no. 12. London, UK: compounds: FEMS Symposium no. 12. London, UK: X Swiss Society of Microbiology on behalf of the Federation of European Microbiological Societies
131. Martins MA, Cardoso MH, Queiroz MJ, Ramalho MT, Campos AMO (1999) Biodegradation of azo dyes by the yeast *Candida zeylanoides* in batch aerated cultures. Chemosphere 38:2455–2460
132. Ozdemir G, Pazarbasi B, Kocyigit A, Omeroglu EE, Yasa I, Karaboz I (2008) Decolorization of Acid Black 210 by Vibrio harveyi TEMS1 a newly isolated bioluminescent bacterium from Izmir Bay Turkey. World J Microbiol Biotechnol 24:1375–1381
133. Guo JB, Zhou JT, Wang D et al (2007) Biocatalyst effects of immobilized anthraquinone on the anaerobic reduction of azo dyes by the salt-tolerant bacteria. Wat Res 41:426–432
134. Kumar K, Devi SS, Krishnamurthi K (2006) Decolorisation, biodegradation and detoxification of benzidine based azo dye. Bioresour Technol 97:407–413
135. Isik M, Sponza DT (2004) Decolorization of azo dyes under batch anaerobic and sequential anaerobic/aerobic conditions. J Env Sci Health 39:1107–1127
136. Panswad T, Iamsamer K, Anotai J (2001) Decolorization of azo reactive dye by polyphosphate and glycogen accumulating organisms in an anaerobic aerobic sequencing batch reactor. Bioresour Technol 76:151–159
137. Xingzu W, Xiang C, Dezhi S, Hong Q (2008) Biodecolorization and partial mineralization of Reactive Black 5 by a strain of *Rhodopseudomonas palustris*. J Env Sci 20:1218–1225
138. Kalyani DC, Patil PS, Jadhav JP, Govindwar SP (2008) Biodegradation of reactive textile dye Red BLI by an isolated bacterium *Pseudomonas* sp. SUK1. Bioresour Technol 99:4635–4841
139. Adedayo O, Javadpour S, Taylor C, Anderson WA, Moo-Young M (2004) Decolorization and detoxification of methyl red by aerobic bacteria from a wastewater treatment plant. World J Microbiol Biotechnol 20:545–550
140. Coughlin MF, Kinkle BK, Bishop PL (1999) Degradation of azo dyes containing amino naphthol by *Sphingomonas* sp. strain ICX. J Industr Microbiol Biotechnol 23:341–346
141. Goldstein RM, Mallory LM, Alexander M (1985) Reasons for possible failure of inoculation to enhance biodegradation. Appl Env Microbiol 50:977–983
142. Bouchez T, Patureau D, Dabert P et al (2000) Ecological study of a bioaugmentation failure. Env Microbiol 2:179–190
143. Marriott CA, Fothergill E, Jeangros B, Scotton M, Louault F (2004) Long term impacts of extensification of grassland management on biodiversity and productivity in upland areas. Agronomie 24:447–462
144. Guieysse B, Wikstrom P, Forsman M, Mattiasson B (2001) Biomonitoring of continuous microbial community adaptation towards more efficient phenol-degradation in a fed-batch bioreactor. Appl Microbiol Biotechnol 56:780–787

145. Kaplan CW, Kitts CL (2004) Bacterial succession in a petroleum land treatment unit. Appl Env Microbiol 70:1777–1786
146. Xia X, Bollinger J, Orgam A (1995) Molecular genetic analysis of the response of three soil microbial communities to the application of 2, 4-D. Mol Ecol 4:17–28
147. Zeng J, Yang L, Du H, Xiao L, Jiang L, Wu J, Wang X (2009) Bacterioplankton community structure in a eutrophic lake in relation to water chemistry. World J Microbiol Biotechnol 25:763–772
148. Liu BB, Zhang F, Feng XX (2006) *Thauera* and *Azoarcus* as functionally important genera in a denitrifying quinoline-removal bioreactor as revealed by microbial community structure comparison. FEMS Microbiol Ecol 55:274–286
149. Watanabe K, Teramoto M, Futamatta H, Harayama S (1998) Molecular detection, isolation and physiological characterization of functionally dominant phenol-degrading bacteria in activated sludge. Appl Env Microbiol 64:4396–4402
150. Daims H, Taylor MW, Wagner M (2006) Wastewater treatment: a model system for microbial ecology. Trends Biotechnol 24:483–489
151. Dabert P, Fleurat-Lessard A, Mounier E et al (2001) Monitoring of the microbial community of a sequencing batch reactor bioaugmented to improve its phosphorus removal capabilities. Wat Sci Technol 43:1–3
152. Eberl L, Schulze R, Ammendola A, Geisenberger O, Erhart R, Sternberg C, Molin S, Amann R (1997) Use of green fluorescent protein as a marker for ecological studies of activated sludge communities. FEMS Microbiol Lett 149:77–83
153. Oerther DB, Danalewich J, Dulekgurgen E (1998) Bioaugmentation of sequencing batch reactors for biological phosphorus removal: comparative rRNA sequence analysis and hybridization with oligonucleotide probes. Wat Sci Technol 37:469–473
154. Urakawa H, Noble PA, El-Fantroussi S, Kelly JJ, Stahi DA (2002) Single base pair discrimination of terminal mismatches by using oligonucleotide microarrays and neutral network analyses. Appl Environ Microbiol 68:235–244
155. Wagner M, Loy A (2002) Bacterial community composition and function in sewage treatment systems. Curr Opin Biotechnol 13:218–227
156. Rittmann BE (2006) Microbial ecology to manage processes in environmental biotechnology. Trends Biotechnol 24:261–266
157. Zhao XQ, Yang LY, Yu ZY et al (2008) Characterization of depth-related microbial communities in lake sediment by denaturing gradient gel electrophoresis of amplified 16S rRNA fragments. J Env Sci 20:224–230
158. Zheng XH, Xiao L, Ren J, Yang LY (2008) The effect of a *Microcystis aeruginosa* bloom on the bacterioplankton community composition of Lake Xuanwu. J Freshwater Ecol 23:297–304
159. Abraham TE, Senan RC, Shaffiqu TS, Roy JJ, Poulose TP, Thomas PP (2003) Bioremediation of textile azo dyes by an aerobic bacterial consortium using a rotating biological contactor. Biotechnol Prog 19:1372–1376
160. Boon N, Top EM, Verstraete W, Siciliano SD (2003) Bioaugmentation as a tool to protect the structure and function of an activated-sludge microbial community against a 3-chloroaniline shock load. Appl Env Microbiol 69:1511–1520
161. Eschenhagena M, Schupplerb M, Oskea IR (2003) Molecular characterization of the microbial community structure in two activated sludge systems for the advanced treatment of domestic effluents. Wat Res 37:3224–3232
162. Niu SQ, Fukushima J, Jiang Y, Ishikawa Y, Ueda T, Matsumoto S (2006) Analysis of bacterial community structure in the natural circulation system wastewater bioreactor by using a 16S rRNA gene clone library. Microbiol Immunol 50:937–950
163. Padayachee P, Ismail A, Bux F (2006) Elucidation of the microbial community structure within a laboratory-scale activated sludge process using molecular techniques. Wat SA 32:679–686
164. Weber SD, Ludwig W, Schleifer KH, Fried J (2007) Microbial composition and structure of aerobic granular sewage biofilms. Appl Env Microbiol 73:6233–6240

165. Yan X, Xu Z, Feng X, Liu Y, Liu B, Zhang X, Zhu C, Zhao L (2007) Cloning of environmental genomic fragments as physical markers for monitoring microbial populations in coking wastewater treatment system. Microbial Ecol 53:163–172
166. Yan X, Xu Z, Feng X et al (2007) Cloning of environmental genomic fragments as physical markers for monitoring microbial populations in coking wastewater treatment system 53:163–172
167. Wakelin SA, Colloff MJ, Kookana RS (2008) Assessing the effect of wastewater treatment pLANT effluent on microbial function and community structure in the sediment of a freshwater stream with variable seasonal flow. Appl Env Microbiol 74:2659–2668
168. Khelifi E, Bouallagui H, Touhami Y, Godon JJ, Hamdi M (2009) Bacterial monitoring by molecular tools of a continuous stirred tank reactor treating textile wastewater. Bioresour Technol 100:629–633
169. Kragelund C, Nielsen JL, Thomsen TR, Nielsen PH (2005) Ecophysiology of the filamentous Alphaproteobacterium *Meganema perideroedes* in activated sludge. FEMS Microbiol Ecol 54:111–122
170. Martins AM, Pagilla K, Heijnen JJ, van Loosdrecht MCM (2004) Filamentous bulking sludge-a critical review. Wat Res 38:793–817
171. Nielsen PH, Roslev P, Dueholm TE, Nielson JL (2002) Microthrix parvicella, a specialized lipid consumer in anaerobic-aerobic activated sludge plants. Wat Sci Technol 46:73–80
172. Fu L, Wen X, Lu Q, Qian Y (2001) Treatment of dyeing wastewater in two SBR systems. Process Biochem 36:1111–1118
173. Kardi F, Eker S, Uygur A (2005) Biological treatment of synthetic wastewater containing 2, 4- dichlorophenol(DCP) in an activated sludge unit. J Env Manage 76:191–196
174. Castillo L, El Khorassani H, Trebuchon P, Thomas O (1999) UV treatability test for chemical and petrochemical wastewater. Wat Sci Tech 39:17–23
175. FitzGerald SW, Bishop PL (1995) Two stage anaerobic/aerobic treatment of sulfonated azo dyes. J Env Sci Health 30:1251–1276
176. Carvalho MC, Pereira C, Gonc-alves IC, Pinheiro HM, Santos AR, Lopes A, Ferra MI (2008) Assessment of the biodegradability of a monosulfonated azo dye and aromatic amines. Int Biodeter Biodegr 62:96–103
177. Ong SA, Toorisaka E, Hirata M, Hano T (2006) Decolorization behavior of azo dye with various co-substrate dosages under granular activated carbon-biofilm configured packed column operation. ARPN J Engin Appl Sci 1:29–34
178. Coughlin MF, Kinkle BK, Bishop PL (2002) Degradation of Acid Orange 7 in an aerobic biofilm. Chemosphere 46:11–19
179. Bras R, Ferra MIA, Pinheiro HM, Goncalves IC (2001) Batch test for assessing decolorization of azo dyes by methanogenic and mixed cultures. J Biotechnol 89:155–162
180. Goncalves IMC, Gomes A, Bras R et al (2000) Biological treatment of effluent containing textile dyes. J Soc Dyers Colourists 116:393–397
181. Sponza DT, Isk M (2004) Decolorization and inhibition kinetic of Direct Black 38 azo dye with granulated anaerobic sludge. Microbial Technol 34:147–154
182. Plumb JJ, Bell J, Stuckey DC (2001) Microbial populations associated with treatment of an industrial dye effluent in an anaerobic baffled reactor. Appl Env Microbiol 67:3226–3235
183. Haug W, Schmidt A, Nortemann B (1991) Mineralization of the sulphonated azo dye mordant yellow 3 by a 6-aminonaphthatene-2-sulphonate-degrading bacterium consortium. Appl Env Microbiol 57:3144–3149
184. Rajaguru P, Kalaiselvi K, Palanivel M, Subburam V (2000) Biodegradation of azo dyes in a sequential anaerobic-aerobic system. Appl Microbiol Biotechnol 54:268–273
185. Togo CA, Mutambanengwe CCZ, Whiteley CG (2008) Decolourisation and degradation of textile dyes using a sulphate reducing bacteria (SRB)–biodigester microflora co-culture. African J Biotechnol 7:114–121
186. Yoo ES, Libra J, Adrian L (2001) Mechanism of decolorization of azo dyes in anaerobic mixed culture. J Env Engin 127:844–849

187. Oxspring DA, McMullan G, Smyth WF, Marchant R (1996) Decolorization and metabolism of the reactive textile dye Remazol-Black-B by an immobilized microbial consortium. Biotechnol Lett 18:527–530
188. Lourenco ND, Novais JM, Pinheiro HM (2000) Reactive textile dye colour removal in a sequencing batch reactor. Wat Sci Technol 42:321–328
189. Lourenco ND, Novais JM, Pinheiro HM (2001) Effect of some operational parameters on textile dye biodegradation in a sequential batch reactor. J Biotechnol 89:163–174
190. O'Neill C, Hawks FR, Hawks DL (1999) Colour in textile effluents-sources, measurement, discharge consents and simulation: a review. J Chem Technol Biotechnol 74:1009–1018
191. Banat IM, Nigam P, Singh D, Marchant R (1996) Microbial decolorization of textile-dye-containing effluents. Bioresour Technol 58:217–227
192. Mendez-Paz D, Omil F, Lema JM (2005) Anaerobic treatment of azo dye Acid Orange 7 under fed-batch and continuous conditions. Wat Res 39:771–778
193. Santos AB (2005) Reductive decolourisation of dyes by thermophilic anaerobic granular sludge. PhD Thesis, Wageningen University, Wageningen, The Netherlands
194. Uddin MS, Zhou J, Qu Y, Guo J, Wang P, Zhao L (2007) Biodecolorization of azo dye Acid Red B under high salinity condition. Bull Environ Contam Toxicol 79:440–444
195. Rieger PG, Meier HM, Gerle M, Vogt U, Groth T, Knackmuss HJ (2002) Xenobiotics in the enviornment: present and future strategies to obviate the problem of biological persistence. J Biotechnol 94:101–123
196. Snyderwine EG, Sinha R, Felton JS, Ferguson LR (2002) Highlights of the eighth international conference on carcinogenic mutagenic N-substituted aryl compounds. Mut Res 506–507:1–8
197. van der Zee FP, Villaverde S (2005) Combined anaerobic aerobic treatment of azo dyes a short review of bioreactor studies. Wat Res 39:1425–1440
198. Carliell CM, Barclay N, Buckley CA (1996) Treatment of exhausted reactive dye bath effluent using anaerobic digestion laboratory and full-scale trials. Wat SA 22:225–233
199. Kapdan IK, Alparslan S (2005) Application of anaerobic–aerobic sequential treatment system to real textile wastewater for color and COD removal. Enz Microbial Technol 36:273–279
200. Kargi F, Uygur A (1996) Biological treatment of saline wastewater in an aerated percolator unit utilizing halophilic bacteria. Env Technol 17:325–330
201. Karigi F, Dincer AR (1998) Saline wastewater treatment by halophile supplemented activated sludge culture in an aerated rotating biodisc contactor. Enz Microbial Technol 122:427–433
202. Lee YH, Pavlostathis SG (2004) Reuse of textile reactive azo dyebaths following biological decolorization. Wat Env Res 76:56–66
203. Lee YH, Matthews RD, Pavlostathis SG (2005) Anaerobic biodecolorization of textile reactive anthraquinone and phthalocyanine dyebaths under hypersaline conditions. Wat Sci Technol 52:377–383
204. Panswad T, Anan C (1999) Impact of high chloride wastewater on an anaerobic/anoxic/aerobic process with and without inoculation of chloride acclimated seeds. Wat Res 33:1165–1172
205. Woolard CR, Irvine RL (1995) Treatment of hypersaline wastewater in the sequencing batch reactor. Wat Res 29:1159–1168
206. Park D, Leeb DS, Jounga JY, Park JM (2005) Comparison of different bioreactor systems for indirect H2S removal using iron-oxidizing bacteria. Process Biochem 40:1461–1467
207. Feng XH, Ou LT, Ogram A (1997) Plasmid-mediated mineralization of carbofuran by *Sphingomonas* sp. strain CF06. Appl Env Microbiol 63:1332–1337
208. Kariminaai-Hamedaani HR, Kanda K, Kato F (2003) Wastewater trewatment with bacteria immobilized onto a ceramic carrier in an aerated system. J Biosci Bioengin 2:128–132
209. Martin M, Mengs G, Plaza E, Garbi C, Sanchez M, Gibello A, Gutierrez F, Ferrer E (2000) Propachlor removal by *Pseudomonas strain* GCHII in an immobilized cell system. Appl Env Microbiol 66:1190–1194

210. Park YS, Yun JW, Kim DS, Song SK (1998) Wastewater treatment in packed bed reacter with immobilized cells onto ceramic carriers. Biotechnol Tech 12:459–462
211. Peres CM, van Aken B, Naveau H, Agathos SN (1999) Continuous degradation of mixtures of 4- nitrobenzoate and 4- aminobenzoate by immobilized cells of *Bulkholderia cepacia* strain PB4. Appl Micobiol Biotechnol 52:440–445
212. Xu P, Quian XM, Wang YX, Xu YB (1996) Modelling for waste water treatment *Rhodopseudomonas palastris* Y6 immobilized on fiber in a columnar bioreacter. Appl Microbiol Biotechnol 44:676–682
213. Guimarães C, Porto P, Oliveira R, Mota M (2005) Continuous decolourization of a sugar refinery wastewater in a modified rotating biological contactor with phanerochaete chrysosporium immobilized on polyurethane foam disks. Process Biochem 40:535–540
214. Hadjiev D, Dimitrov D, Martinov M, Sire O (2007) Enhancement of the biofilm formation on polymeric supports by surface conditioning. Enz Microbial Technol 40:840–848
215. Zaiat M, Cabral AKA, Foresti E (1996) Cell washout and external mass transfer risistance in horizontal-flow anaerobic immobilized sludge reactor. Wat Res 30:2435–2439
216. Nojiri H, Shintani M, Omori T (2004) Divergence of mobile genetic elements involved in the distribution of xenobiotic-catabolic capacity. Appl Microbiol Biotechnol 64:154–174
217. Sørensen SJ, Bailey M, Hansen LH, Kroer N, Wuertz S (2005) Studying plasmid horizontal transfer in situ. Nature Rev Microbiol 3:700–710
218. Top EM, Springael D (2003) The role of mobile genetic elements in bacterial adaptation to xenobiotic organic compounds Curr Opin Biotechnol 14:262–269
219. Springael D, Top EM (2004) Horizontal gene transfer and microbial adaptation to xenobiotics: new types of mobile genetic elements and lessons from ecological studies. Trends Microbiol 12:53–58
220. Top EM, Springael D, Boon N (2002) Catabolic mobile genetic elements and their potential use in bioremediation of polluted soils and waters. FEMS Microbiol Ecol 42:199–208
221. van der Meer JR, Sentchilo V (2003) Genomic islands and the evolution of catabolic pathways in bacteria. Curr Opin Biotechnol 14:248–254
222. van Elsas JD, Bailey MJ (2002) The ecology of transfer of mobile genetic elements. FEMS Microbiol Ecol 42:187–197
223. van Elsas JD, Duarte GF, Keijzer WA, Smit E (2000) Analysis of dynamics of fungal specific PCR of soil DNA followed by denaturing gradient gel electrophoresis. J Microbiol Methods 43:133–151

Biodegradation of Azo Dyes Under Anaerobic Condition: Role of Azoreductase

S. Sandhya

Abstract The pressures of an ever-increasing population and industrial development have led to the addition of an array of man-made chemicals in the environment, leading to a tremendous deterioration in environmental quality. Contamination of soil, air, water, and food is one of the major problems facing the industrialized world today. Significant regulatory steps have been taken to eliminate or to reduce the production or release of these chemicals into the environment. A major contribution to these categories is by azo dyes, most of which are toxic and hazardous in nature. Application of microbial processes to decontaminate environmental media polluted with these compounds will require a better understanding of why and how microorganisms can degrade them and utilize them for their own survival as well as for cleaning the environment. This review focuses on different anaerobic microbial processes for biodegradation of azo dyes and enzymes involved therein that are responsible for their degradation.

Keywords Anaerobic, Azo dyes, Biodegradation

Contents

S. Sandhya
National Environmental Engineering Research Institute, CSIR Complex, Taramani, Chennai 600113, India
e-mail: s_sandhya@neeri.res.in

H. Atacag Erkurt (ed.), *Biodegradation of Azo Dyes*,
Hdb Env Chem (2010) 9: 39–57, DOI 10.1007/698_2009_43,

Abbreviations

CI Color index
SRB Sulfate reducing bacteria

1 Introduction

Industrialization is considered to be the key to development in economic terms. However, it is also recognized to be the root problems from environmental perspective. The recognition that environmental pollution is a worldwide threat to public health has given rise to new initiatives for environmental restoration for both economic and ecological reasons. The industrial effluents contain toxic and hazardous pollutants. One particular class of synthetic chemicals which is of major concern is synthetic dyes and dye intermediates. The dyes are extensively used for textile, paper printing, and color photography, cosmetic, pharmaceutical, and leather industries. In the 1994 estimates, the world production of dyes was around 1 million tons, of which more than 50% were azo dyes [1, 2]. India, Eastern European countries including the USSR, China, South Korea, and Taiwan together consume approximately 600,000 tons of dyes per annum [3]. Even though the dye industry is characterized by a large number of producers (about 2,000 world wide), just four western companies accounted for nearly half of the market in 2000 [4]. It is estimated that more than 50% of these dyes are lost into wastewater, depending on the class of dyes used [5]. Azo dyes, which are aromatic compounds with one or more –N=N– groups, constitute the largest class of synthetic dyes in commercial applications [6]. Dyes can be toxic and mutagenic, and if they are discharged directly into the environment, they persist as environmental pollutant as well as traverse through the entire food chains, leading to biomagnifications. Many dyes are visible in water at concentration as low as 1 mg L^{-1}. Dyes that are absorbing light with wavelength in visible range (350–700 nm) are colored. Dyes contain chromophores electron systems with conjugated double bonds and auxochromes, electron withdrawing or electron donating substituents that cause the color of the chromophore by altering the overall energy of the electron system. Usual chromophores are –C=C–, –C=N–, –C=O, –N=N–, $-NO_2$, and quinoid rings, and usual auxochromes are $-NH_3$, –COOH, $-SO_3H$, and –OH.

1.1 Classification of Dyes

Dyes are classified in accordance with either the chemical constitute or their application to textile fibers for coloring purposes. Table 1 gives this classification

Table 1 Classification of dyes according to usage

S No.	Class	Major substrates	Method of application	Remark
1	Acid	Nylon, wool, silk, paper, inks, and leather	Usually form neutral to acidic dye baths	This group of dyes is very important for wool of protein fibers. The important premetallized dyes are members of this class
2	Azoic dyes	Cotton, rayon, cellulose acetate, polyester paper, and inks	Fiber impregnated with coupling component and treated with a solution of stabilized dia-azonium salt	Dyeing and printing cotton in fast shades
3	Basic	Acrylic, modified nylon and polyester paper, and inks	Applied from acidic dye baths	Known as cationic dyes yield intense brilliant shades but fugitive to light
4	Direct	Cotton, rayon, paper, leather, and nylon	Applied from neutral or slightly alkaline baths containing additional electrolyte	Very important class of dyes
5	Disperse	Polyester polyamide, cellulose acetate, acrylic, and plastics	Fine, aqueous dispersions often applied by higher temperature, pressure, or lower temperature carrier methods, dye may be padded on cloth, baked on or thermo-fixed	New fast growing field of dyes important for synthetic fibers
6	Optical brighteners	Soap and detergents, all fibers, oils, paints, and plastics	From solution dispersion or suspension in a mass	
7	Reactive	Cotton, rayon, wool, silk, and nylon	Reactive site on dye reacts with functional group on the fiber to bind dye covalently under influence of heat and proper pH	Bonds chemically to the fiber
8	Sulfur	Cotton and rayon	Aromatic substrate valued with sodium sulfide and reoxidized to insoluble sulfur containing products on the fiber	Sulfur black belongs to this group
9	Vat	Cotton, rayon, and wool	Water insoluble dyes solublized by reducing with sodium hydro sulfite then exhausted on fiber and reoxidized	Characterized high fastness specially anthraquinoids – most valuable for dyeing and printing cotton

(continued)

Table 1 (continued)

S No.	Class	Major substrates	Method of application	Remark
10	Solubilized vat	Cotton, wool, cellulose, and protein fibers silk	Impregnated fiber when treated with an oxidized agent usually sulfuric acid and sodium nitrite for cotton dichromate wool and silk; therefore, no alkali is involved. This class is applicable to cellulose and protein fibers	Dyeing, printing and wool in fast shades

of dyes according to usage and Table 2 according to chromophore. The dyes are anionic (direct, acid, and reactive dyes), cationic (basic dyes), and nonionic dispersed dyes. Anionic and nonionic dyes mostly contain azo or anthroquinone type of chromophores. Table 3 gives the classification of dyes according to their properties.

The dyes used in olden days were natural dyestuffs such as Saffron, Henna, Cochineal, Logwood, etc. derived from plants or animals. Water extracts of various plants or animals gave solution of yellow, and browns from the extract of the Mediterranean mollusk, and indigo came from plants of the genus *Indigofera*. Most natural dyes are of mordant type that requires a fixing agent. Amongst the synthetic dyes used in these industries, azo dyes are the oldest and play a prominent role in almost every type of application [7]. A substituent often found in azo dyes is the sulfonic acid group ($-SO_3H$) called sulfonated azo dyes.

1.2 Color Index

The color index (CI) number, developed by the society of dyers and colorists, is used for dye classification. Once the chemical structure of a dye is known, a five-digit CI number is assigned to it. The first word is the dye classification and the second word is the hue or shade of the dye. For example, CI Acid Yellow 36 (CI 13065) is a yellow dye of the acid type. Additionally, a dye mixture may consist of several dyes; for example, Navy 106 is composed of three reactive azo dyes: remazol black B (Reactive Black 5), Remazol Red RB (Reactive Red 198), and Remazol Golden Yellow 3.

1.3 Color Measurements

Qualitatively, the type of the color and its intensity are easily visualized through our eyes. The peak absorbance of a given wave length (λ_{max}) for a particular color is

Table 2 Classification of synthetic organic colorants by chromophores

Chromophore	Dye class	Chromophore	Dye class
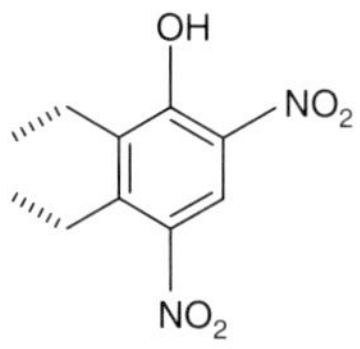	Nitro	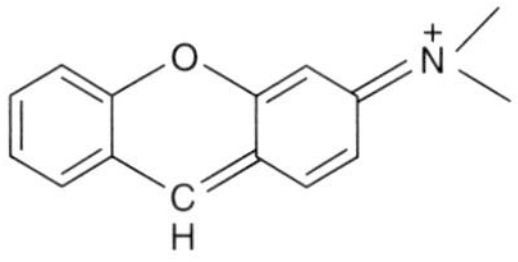	Xanthenes
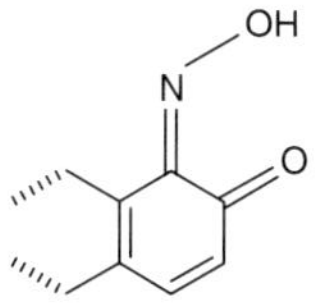	Nitroso	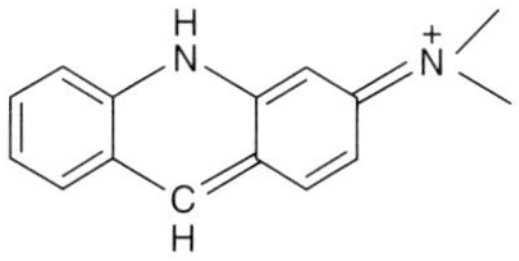	Acridines
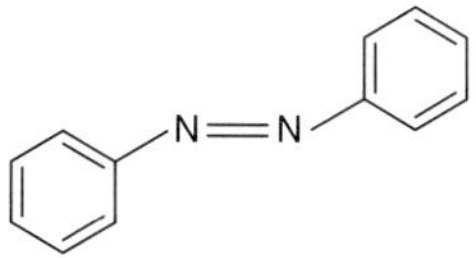	Azo	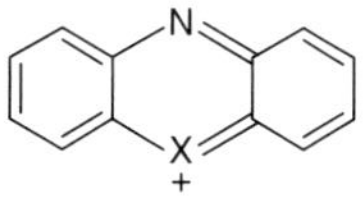	Quinoneimines
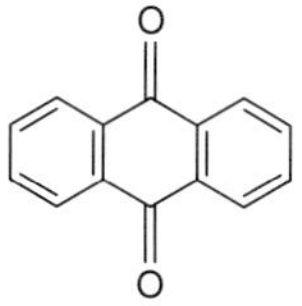	Anthraquinones	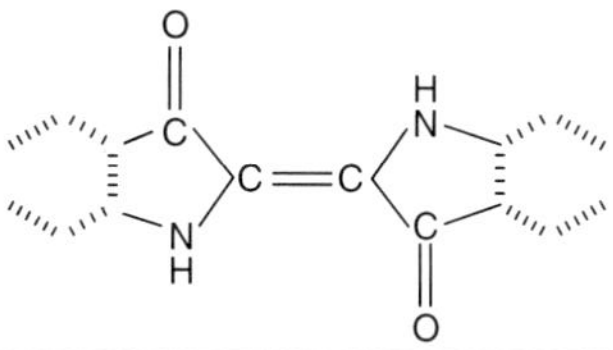	Indigoids

(*continued*)

Table 2 (continued)

Chromophore	Dye class	Chromophore	Dye class
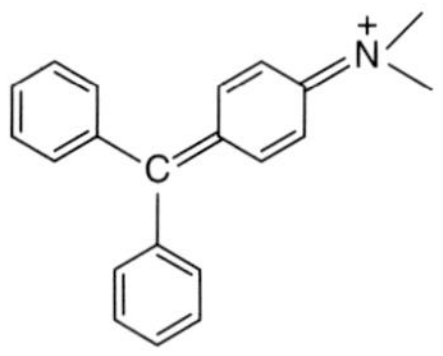	Triphenylmethanes	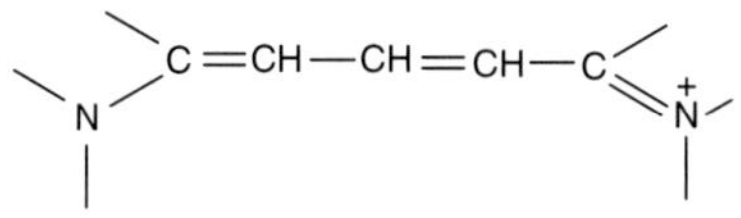	Cyanines
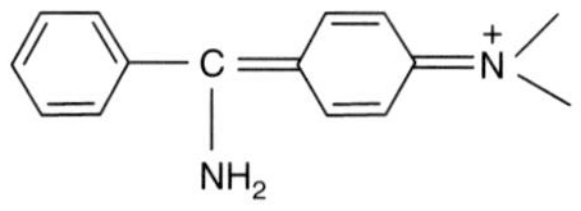	Diphenylmethanes		

Table 3 Classification of azo dyes by properties

Azo dye	Dye class and substrate	Dominant mechanism of binding
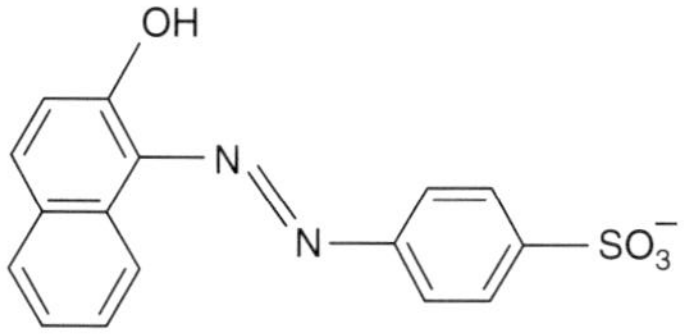	Anionic dye for wool	Electrostatic
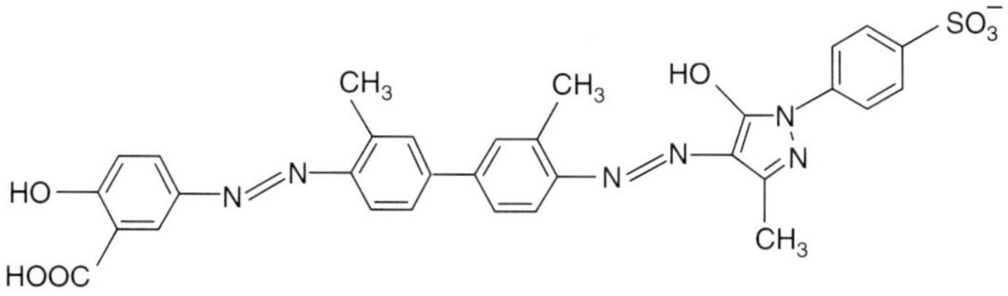	Direct dye for cotton	Vander waal
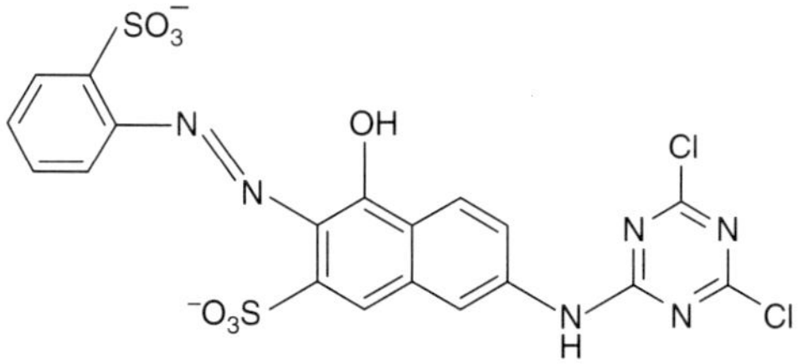	Reactive dye for cotton	Chemical reaction
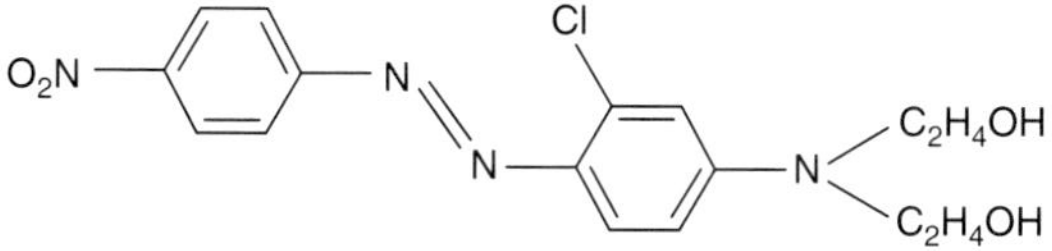	Dispersed dye for polyester	Solubility

Table 4 Relationship between wavelength and color

Wavelength (nm)	Perceived hue
400–440	Green–yellow
440–480	Yellow
480–510	Orange
510–540	Red
540–570	Purple
570–580	Blue
580–610	Greenish-blue
610–670	Blue–green

easy to measure in a visible spectroscopy to quantify the magnitude of absorbance to the color intensity. The range of the wavelength corresponding to a particular color is shown in Table 4. Currently there are five methods for determining the color of the samples in the latest edition of the standard methods: visual comparison, spectrophotometric, tristimulus, ADMI methods [8]. The visual method is based on the sample comparison with a standard color solution of platinum cobalt chloride. This method is applicable to the sampling of potable water, but is inadequate for quantifying the color intensity in dye wastes. All these methods generally measured color caused by stable compounds; hence color value may be underestimated due to colloidal colored particles.

Biological methods are currently viewed as effective, specific, less energy intensive, and environmentally benign since they result in partial or complete bioconversion of organic pollutants to stable nontoxic end products. Even though azo compounds are xenobiotic in nature and expected to be recalcitrant to biodegradation. It is known that dyes resist biodegradation in conventional activated sludge treatment units [2]. The number of microorganisms now known, including bacteria, fungi, yeast, and algae, can decolorize the dyes and even completely mineralize many azo dyes under certain environmental conditions. Many reviews are available on microbial methods of decolorization of azo dyes [2, 9–18]. Table 5 gives examples of some common dyes used in microbial degradation study. Biodegradation processes may be anaerobic, aerobic, or involve a combination of the two. When considering the reaction between bacterial cells and azo dyes, it must be noted that there are significant differences between the physiologies of microorganisms grown under aerobic and anaerobic conditions [2]. For aerobic bacteria to be significant in the reductive process, the bacteria must be specifically adapted. This adaptation involves long-term aerobic growth in continuous culture in the presence of a very simple azo compound. The bacteria synthesize an azoreductase specific for this compound which, under controlled conditions, can reductively cleave the azo group in the presence of oxygen [2]. In contrast, bacterial reduction under anaerobic conditions is relatively unspecific with regard to the azo compounds involved, and is, therefore, of more use for the removal of color in azo dye wastewater [2].

Table 5 Dyes used in biodegradation studies

Acid Orange 6

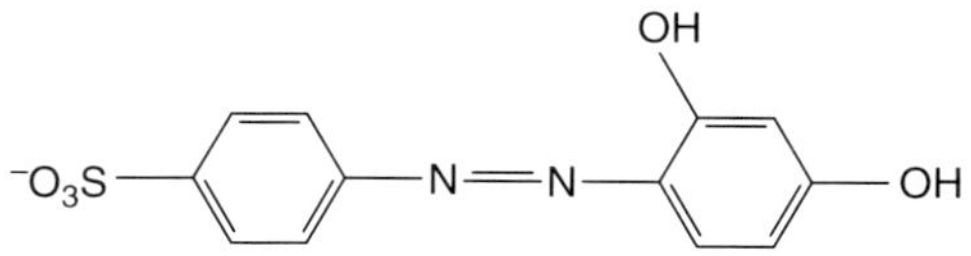

Acid Red 14

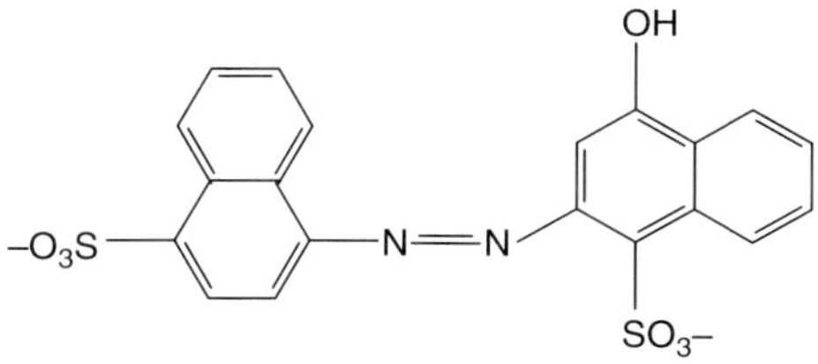

Acid Yellow 9

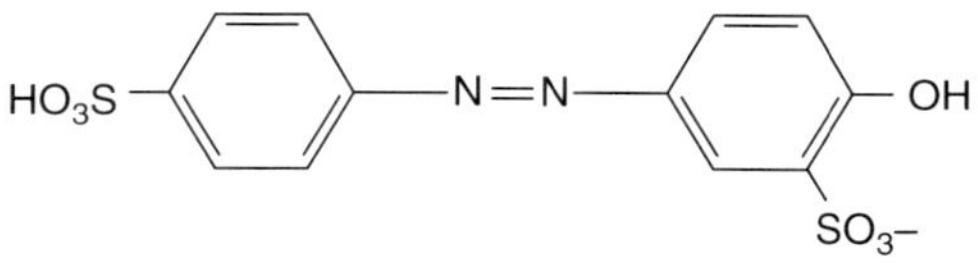

Carboxy Orange 1

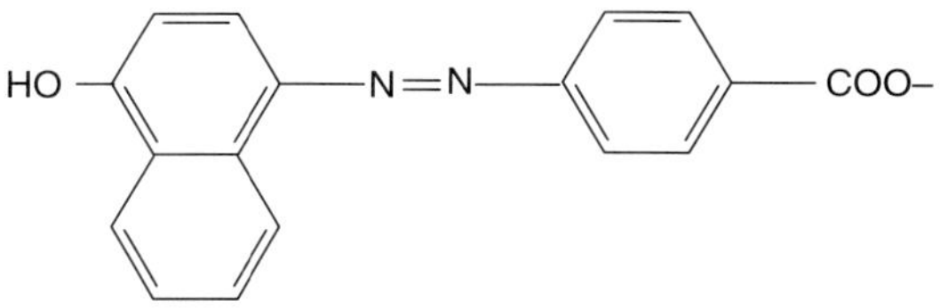

Direct Red 28

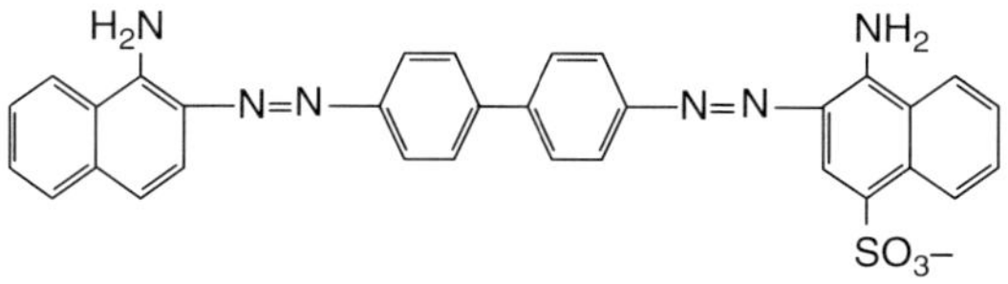

Disperse Yellow

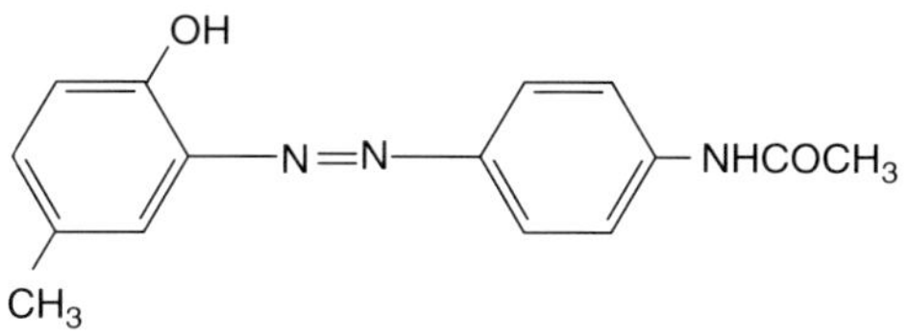

Methyl Red

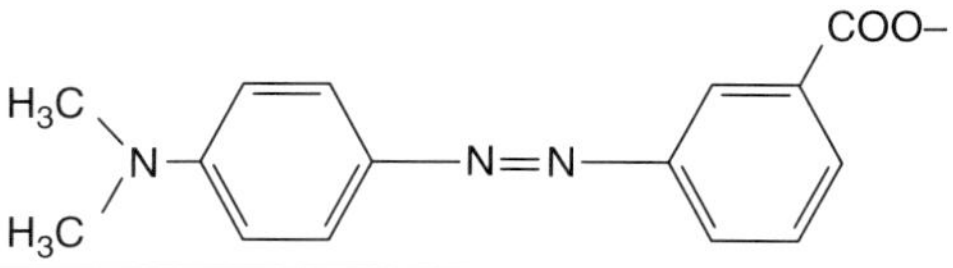

(continued)

Table 5 (continued)

Mordant Yellow 3

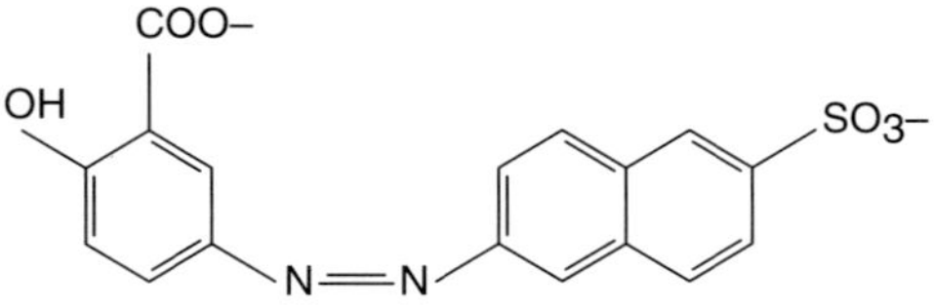

Reactive Violet 5

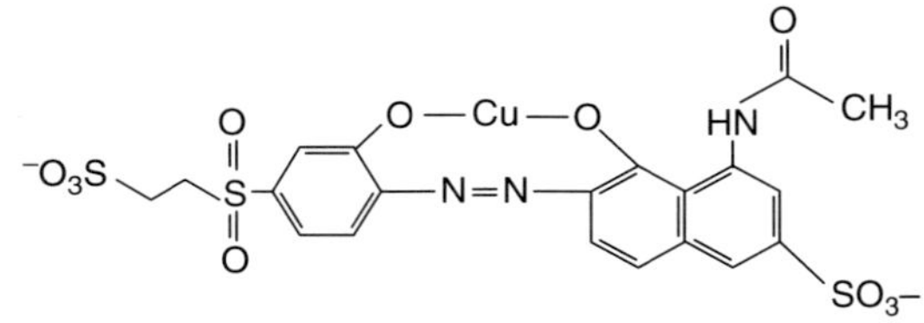

Acid Red G

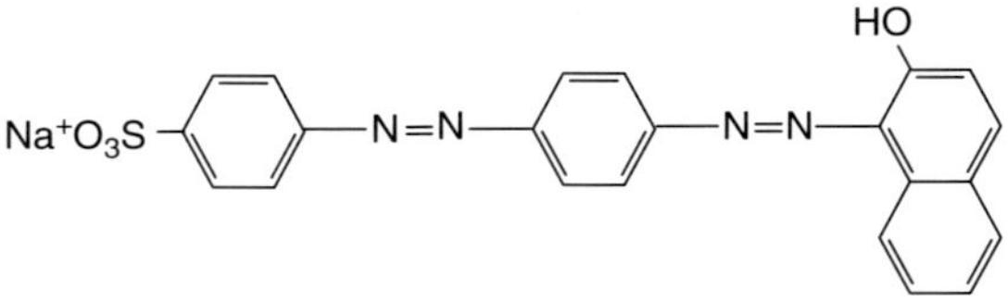

Acid Red 88

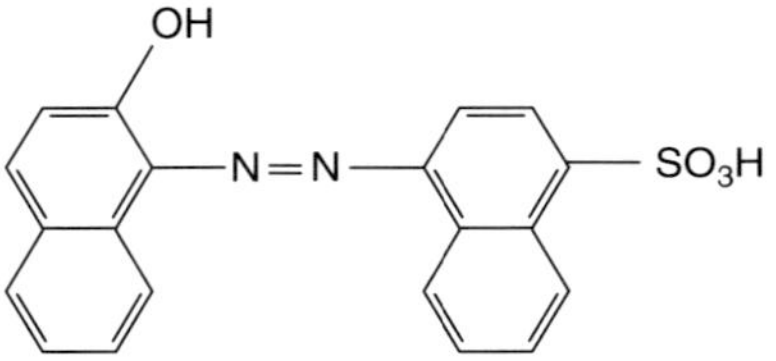

1-(4′-Carboxy-phenylazo)-2-napthol

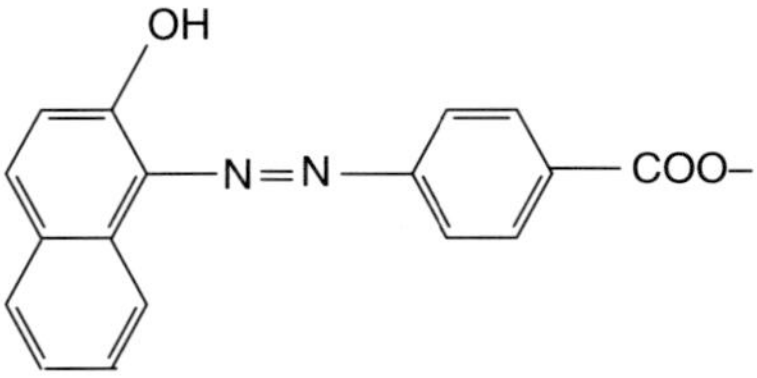

Acid Orange 7

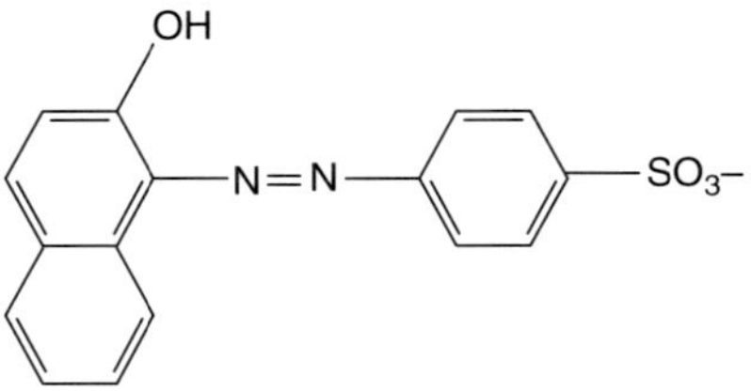

4′-Carboxy-4′-sulfoazo-benzene

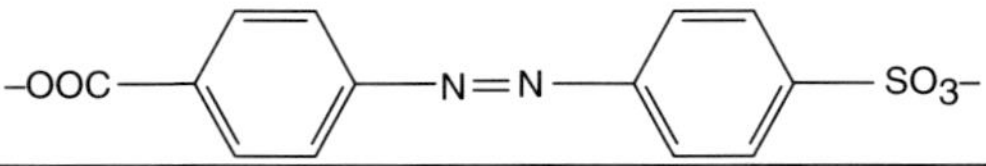

(*continued*)

Table 5 (continued)

4,4′-Dicarboxy-azobenzene	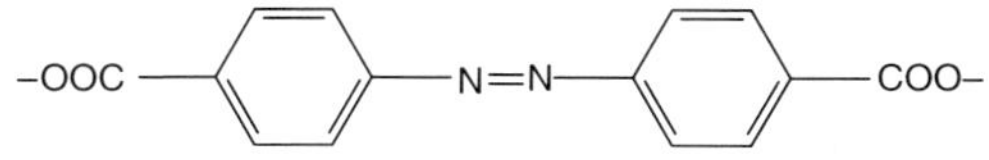
Methyl Red	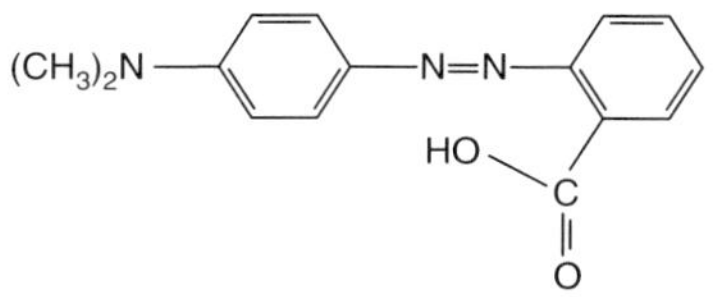
Remazol Black	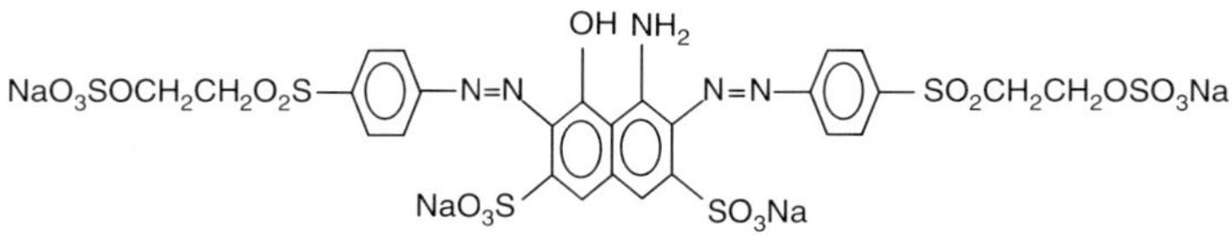
Reactive Orange 16	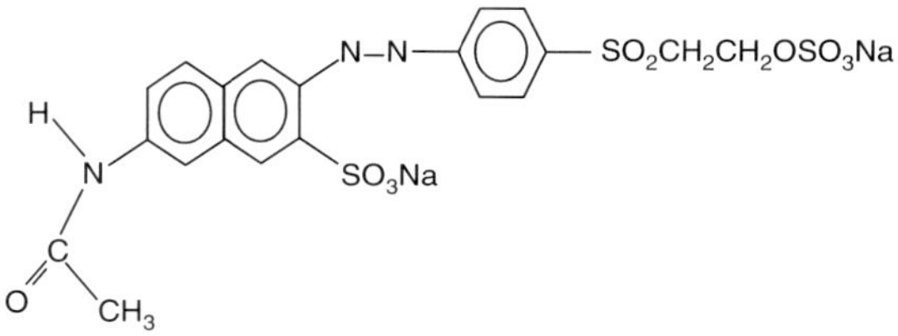
Reactive Red 11	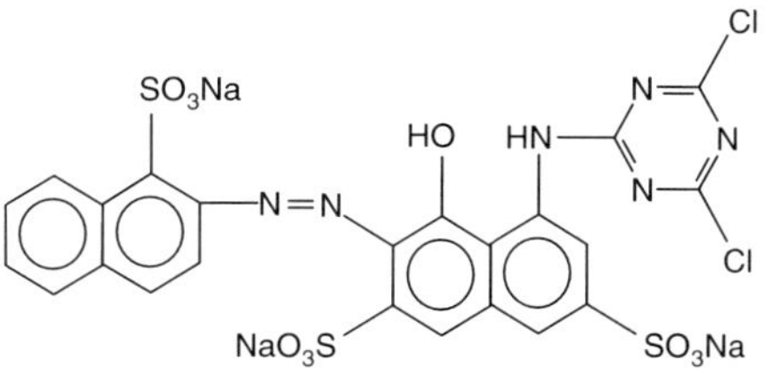
Acid Yellow 36	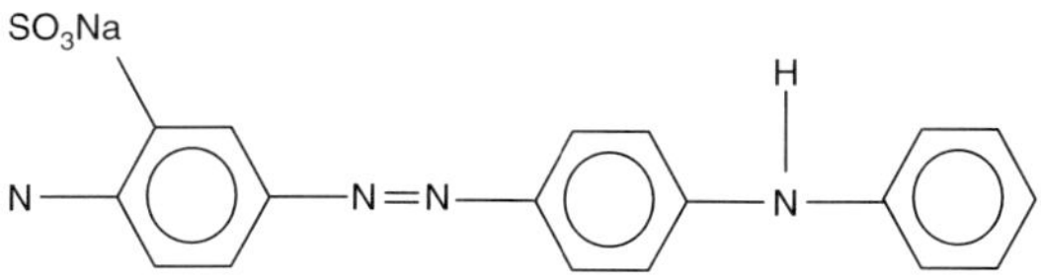
Reactive Red 141	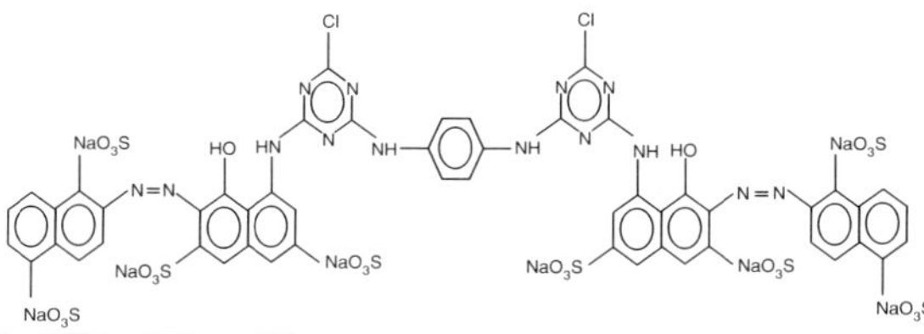

(*continued*)

Table 5 (continued)

Direct Green 6	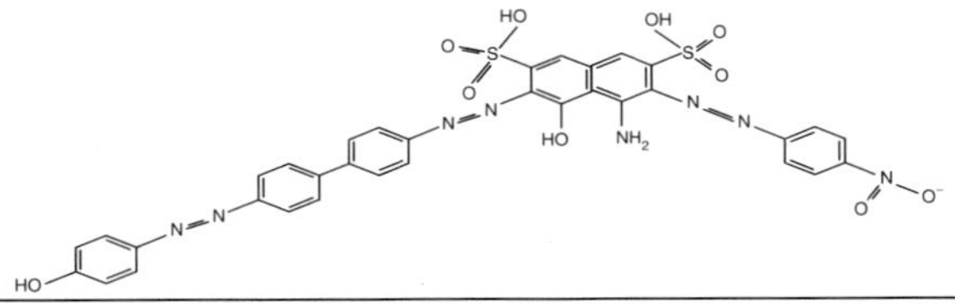

2 Anaerobic Degradation of Dyes

Anaerobic reduction of azo dyes using microbial sludge can be an effective and economic treatment process for removing color from wastewater. The investigation on anaerobic decolorization of azo dyes was started long back in early 1970s. Walker and Rayan reported decolorization of azo dyes using intestinal aerobic bacteria [19]. This potential of intestinal anaerobes to decolorize the azo dyes was further established by other researchers [20–24]. Previous studies have demonstrated the ability of anaerobic bacteria to reductively cleave the azo linkages in reactive dyes. Although this effectively alters the chromogen and destroy the observed colors of the dye, many aromatic groups are not susceptible to anaerobic reduction. Dye decolorization under methanogenesis condition requires an organic carbon/energy source. Simple substrates like glucose, starch, acetate ethanol, whey, and tapioca have been used as dye decolorizing substrate [25–27]. Chinrelkitvanich et al. [25] and Bras et al. [28] have shown that acidogenic as well as methanogenic bacteria contribute to dye decolorization. They have used molecular methods to characterize the microbial population in anaerobic buffered reactor; tracking industrial dye waste showed that members of proteobacteria together with sulfate-reducing bacteria (SRB) were prominent members of mixed bacterial population. The methanogenic population was dominated by *Methanosaeta* species and *Methano methyovorams hollandia* [29]. Yoo et al. showed that the decolorization of Orange 96 was significantly affected by 2 bromo ethane sulfonic acid, an inhibitor specific to methanogens. This suggest that methanogens does not have any part in decolorization [30]. On the other hand, in the presence of acetate, sulfate molybdate inhibits SRB, which has a significant effect on decolorization rate. Reduction under anaerobic conditions appear to be nonspecific, as most varied azo compounds are decolorized, although the rate of decolorization depends on the organic carbon and the dye structure [2]. Some azo dyes are more resistant to removal by bacterial cells [28]. Dyes with simple structures and low molecular weights exhibit higher rates of color removal, whereas color removal is more difficult with highly substituted, high molecular weight dyes [31]. In the case of the terminal nonenzymatic reduction mechanism, reduction rates are influenced by changes in electron density in the region of the azo group. The substitution of electron-withdrawing groups ($-SO_3H$, $-SO_2NH_2$) in the *para* position of the phenyl ring, relative to the azo bond, causes an increase in the reduction rate [19]. Nigam et al. established that azo compounds with a hydroxyl group or with an amino group are more likely to be degraded than those with a methyl, methoxy, sulfo, or nitro groups [32]. Color removal is also related to the number of azo bonds in the dye molecule. The color of mono-azo dyes is removed faster than the color of diazo or triazo dyes. Hu showed that the turnover rate of mono-azo dyes increased with increasing dye concentration, whereas the turnover rate of the diazo dyes and of the triazo dyes remained constant as the dye concentration increased [33]. Hitz et al. concluded that acid dyes exhibit low color removal due to the number of sulfonate groups in the dye, or direct dyes exhibit high levels of color removal that is independent of the number of sulfonate groups in the dye, and reactive dyes exhibit low levels of color removal [34]. The effect of the sulfonate groups on color removal is related to the mechanism by which the color is removed. If the dye reduction takes place inside the cell, the presence of sulfonate groups will hinder the transfer of the dye molecule through the cell membrane. Therefore, the rate of dye reduction will decrease as the number of sulfonate groups increases. However, if the dye reduction takes place outside the cell, the presence of sulfonate groups will have little effect on the rate of dye reduction. Kulla found that cultures could be adapted to produce azoreductase enzymes that had very high specificity towards particular dye structures [35]. One such enzyme, Orange I azoreductase, exclusively reduced the azo groups of Orange I and its derivatives, with their hydroxyl group in the *para* position. Another enzyme, Orange II azoreductase, was specific for Orange II-type compounds, with their hydroxyl group in the *ortho* position. It was also shown that sulfonated dyes were reduced faster than carboxylated dyes due to the higher electronegativity of the sulfo group, which renders the azo group more accessible to electrons. Hydrogen bonding, in addition to the electron density in the region of the azo bond, has a significant effect on the rate of reduction [36]. The position and the nature of substituents on the dye molecule influence the azo-hydrazone tautomerism of hydroxyazo compounds. The hydroxy proton of phenylazo-naphthol derivatives is labile and can bond with a nitrogen atom of the azo group, causing a rapidly formed tautomeric equilibrium between the azo and the hydrozone forms. This equilibrium is influenced by both structural factors within the molecule and by the nature of the medium surrounding the molecule. Zimmerman et al. found that, with certain azoreductases, a decreased rate of reduction was observed when the enzyme system was run with a substrate that was stabilized in the hydrazone form via hydrogen bonding, suggesting that the azo configuration of the substrate molecule was important for the enzymatic reaction [37]. However, the degree of interference caused by the methyl group could not be appraised. Zimmerman et al. made some generalizations with respect to the structural features that are required of the substrates for reduction by bacteria exhibiting the Orange-II azoreductase: (a) a hydroxy group in the *ortho* position of the naphthol ring is a prerequisite for the reaction; (b) charged groups in the proximity of the azo group will hinder the reaction; (c) a second polar substituent on the dye molecule lowers its affinity to the enzyme and inhibits the reaction; and (d) electron withdrawing substituents on the phenyl ring will increase the rate of the reaction [38].

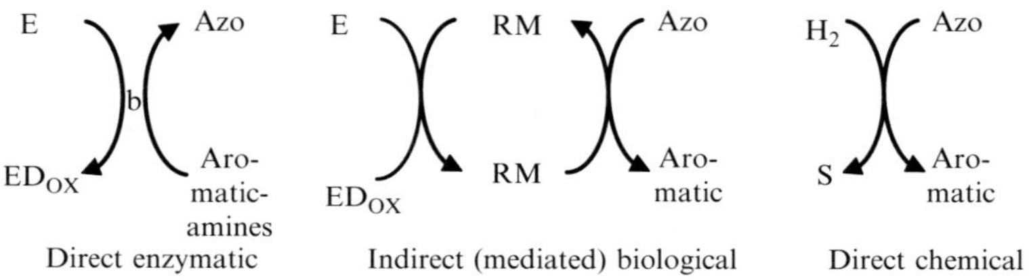

Fig. 1 Mechanism of anaerobic azo dye reduction. *RM* redox mediator; *ED* electron donor; *b* bacteria (enzyme) [14]

Decolorization appears as fortuitous process, where dye acts as electron transport chains. Alternatively, decolorization might be attributed to nonspecific extracellular reaction occurring between reduced compounds generated by the anaerobic biomass [39]. The ubiquitous range of microorganisms that reduce azo compounds under anaerobic condition. Flavin reductase acts like azoreductase in in vitro experiments and may be responsible for unspecific reaction of azo dyes [40]. Under strict anaerobic conditions, decolorization of dyes can be enhanced in the presence of redox mediators such as benyl viologen a quinines [39]. Extracellular reduction of azo dyes by microorganisms may also be due to reduced inorganic compounds such as Fe^{+2} and H_2S, which are formed as anaerobic bacterial metabolic reactions and products. Figure 1 represents the different mechanisms of anaerobic azo dyes reduction. *Clostridium paraputrificans* was found capable of reducing seven commercially available structurally related azo dyes [41]. Mediated biological azo dye reduction as highly polar sulfonated, as well as high molecular weight, polymeric azo dyes are unlikely to pass through the cell membrane [42]. It was suggested that reduction of these dyes could also occur through mechanisms that are not dependent on their transport into the cell. There are now many reports on the role of redox mediators in azo bond reduction by bacteria under anaerobic conditions [39, 43, 44]. Riboflavin in catalytic amounts significantly enhanced the reduction of mordant yellow 10 by anaerobic granular sludge [45]. 1-Amino 2-napthol, one of the constituent amines of the azo dye, AO7, increased its decolorization rate possibly by mediating the transfer of reducing equivalents [22]. The addition of synthetic electron carriers such as anthraquinone-2,6-disulfonate could also greatly enhance the decolorization of many azo dyes [39]. Keck et al. reported the first example of the anaerobic cleavage of azo dyes by redox mediators formed during the aerobic degradation of a xenobiotic compound [44]. Cell suspensions of *Sphingomonas* sp. Strain BN6 grown aerobically in the presence of 2-naphthylsulfonate (NS) exhibited a 10–20-fold increase in decolorization rate of an azo dye, amaranth, under anaerobic conditions over those grown in its absence. Even the addition of culture filtrates from these cells could enhance anaerobic decolorization by cell suspensions grown in the absence of NS. Based on these observations, a mechanism was proposed for the mediated reduction of azo dyes by *S. xenophaga* (Fig. 2). Other bacterial cultures generating redox intermediates during the aerobic

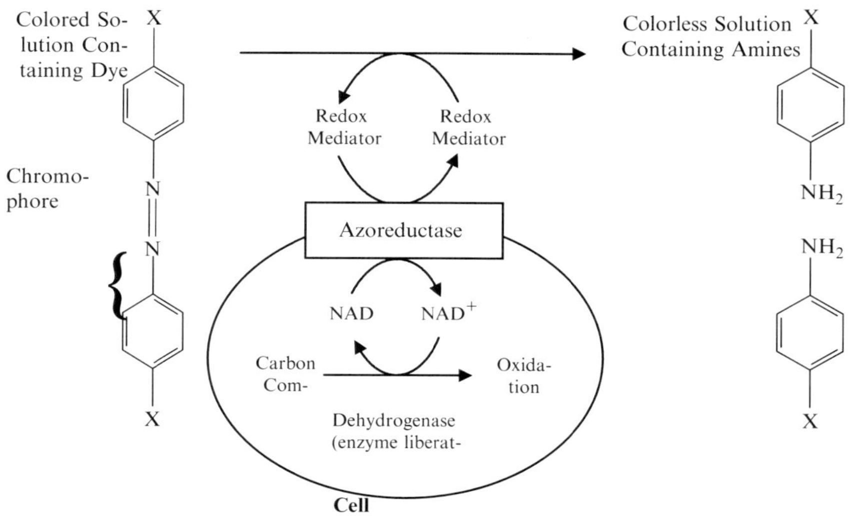

Fig. 2 Proposed mechanism for reduction of azo dyes by bacterial cell [44]

degradation of aromatic compounds can also lead to the enhancement of dye decolorization in anaerobic conditions [44]. In anaerobic treatment azo reduction is achieved by cleaving azo bond, but toxic amines are generated. The problem due to break-down product can be overcome by a sequential or simultaneous two-stage process. In the microbial degradation of azo dyes, the initial process is their decolorization. The highly electrophilic azo bond gets cleaved for azo decolorization. Till the year 2000, it has been reported that neither the activated sludge nor the aerobic bacterial isolates were able to degrade azo dyes [2]. On the other hand, various azo dyes were shown to be developed by anaerobic sludges, anaerobic sediments [42], and pure culture of bacteria incubated anaerobically.

But afterwards it was found that facultative microorganisms behave differently in sequential treatment. Sandhya et al. studied the functions of the microaerophilic–aerobic sequential batch reactor for the treatment of synthetic dye wastewater [46].

3 Role of Azoreductase in Biodegradation

The enzyme involved in the degradation of the dyes has been shown to be azoreductase. The enzymes were first isolated from the intestinal microflora and was later found to be produced by the cytosolic and microsomal fractions of the liver [47]. The enzyme was sensitive to oxygen and was inactivated by oxygen. In experiments involving intestinal anaerobic bacteria, Rafii et al. found the requirement of coenzyme FAD, FMN, or riboflavin for the enzyme activity [47]. The average rate of reduction of the dye by these strains ranged from 16 to 135 nmols of dye per minute per milligram of protein. Nondenaturing polyacrylamide gel electrophoresis showed that each bacterium expressed only one azoreductase isozyme. It was found that at least three types of azoreductase were produced by different isolates. All the azoreductase were produced constitutively and released extracellularly. The extracellular enzyme activity was recorded to be higher than that of the intracellular enzyme source.

The inability of the cultures to utilize the accumulated metabolites in anaerobic degradation of the dyes led to the isolation of strains, which possessed oxygen insensitive azoreductases. Kulla et al. isolated *Pseudomonas* K22 and KF46 strains from chemostate cultures that were adopted to grow on carboxyl orange I and carboxyl orange II, respectively [35]. They were able to degrade sulfonated analogs of carboxyl orange I and carboxyl orange II. Oxygen-insensitive azoreductases have been detected in these organisms, which have been selected by long-time adaptation in chemostate for growth on azo dyes. These sulfonated dyes served as models for aerobic treatment of wastewater from textile industries, where such dyes had wide usage. The degradative pathways for azo dyes were initiated by oxygen insensitive azoreductase, which catalyzed the reductive cleavage of the azo group with NAD(P)H as an electron donor [38] and the dyes as artificial electron acceptors. Under aerobic conditions, the aromatic amines resulting from the fission of carboxylated orange dyes were mineralized. Two enzymes were responsible for the reduction, Orange I and Orange II azoreductases. This study has initiated the occurrence of each azoreductase exclusively in bacteria that were selected for utilization of carboxylated analogs of the primary enzyme substrate as the sole carbon, nitrogen, and energy source. Both enzymes thus played key roles in the degradation of the type of azo dye that was used as a selective agent in experimental evolution in continuous cultures. The common features of the two azoreductases include their monomeric structure, their specificity for NADPH and NADH as cosubstrates, substrate inhibition, and the order of magnitude of the K_m values for primary substrates and for the cosubstrates, as well as the temperature and pH optima of the reaction catalyzed by the enzymes.

The two azoreductases differed strictly in their specificity with regard to the position of the hydroxyl group on the naphthol ring of the substrates. Molecules with none or with two hydroxyl groups were not reduced by the enzymes. Orange I azoreductase exclusively accepted substrates with a hydroxyl group in the $4'$ position of the naphthol moiety, while Orange II azoreductase was specific for orange dyes carrying a hydroxyl group in the $2'$ position [38]. The two enzymes also differed in the molecular weight and regulation of their synthesis. Orange I azoreductase is a constitutive enzyme with a molecular weight of approximately 21,000, whereas Orange II azoreductase was a protein with a molecular weight of 30,000, which was induced by various azo dyes. The enzymes did not exhibit immunological cross reaction with each other, and since cross inhibition by their antisera was not observed, the results of the immunological tests with the pure enzymes argued against a close relationship between the azoreductases.

4 Mechanism of Azo Reduction

The first step in the bacterial degradation of azo dye in either aerobic or anaerobic condition is the reduction of –N=N– bond. Reduction may be due to enzymes, redox mediator, and chemical reduction by reductants like sulfide or combination. This reaction involving enzyme-mediated azo dye reduction may be either specific or nonspecific to dye. The presence of azoreductase in anaerobic bacteria was first reported by Rafii et al. in *Clostridium* and *Eubacterium*. Azoreductase from these strains were oxygen sensitive and were produced constitutively and released extracellularly [47]. Later investigation made by Rafii and Cernglia has shown azo reduction in *Clostridium perfringens* by an enzyme FAD dehydrogenase [48]. The gene for this enzyme for *C. perfringens* has been cloned and expressed in *Escherichia coli* [49]. Another mechanism of dye decolorization could involve cytosolic flavin-dependent reductions, which transfer electron via soluble flavins to azo dyes. However, recently Russ et al. have shown that recombinant *Sphingonomas* strain BN6 could reduce sulfonated azo dyes by cytosolic flavin-dependent azo reduction in vitro and not in vivo [40].

5 Conclusion and Future Research Needs

The fate of synthetic chemicals reaching the environment for the most part depends on the microorganisms present in that part of the environment. The capacity of microbes to produce enzymes that recognize xenobiotic compounds and to catalyze reactions that break them decides the extent to which such chemicals can cause damage to the ecosystem. The absence of microorganisms or microbial systems that bring about their degradation will only result in these chemicals being recalcitrant, persistent, and a potent hazard to the ecosystem as a whole. Microbial degradation of azo dyes in general can be divided into cometabolic conversions or conversions that yield energy and are metabolically productive. Azo dyes over a range act as the sole source of carbon and energy to different groups of bacteria. Since azo dyes are electron deficient, they act as electron acceptors, and energy is generated in a respiratory process in anaerobic environments. In fortuitous metabolism, the azo compound is degraded only in the presence of another substrate and is degraded due to the presence of broad-specificity enzymes in bacteria. Enzymes produced by organisms that degrade azo dye compounds are coded by genes that are chromosomally carried. Many of them are transposable elements. A vast number of such genes have been characterized and strategies for engineered organisms that carry genes for biodegradation have been constructed.

Future research related to biodegradation of azo dyes should focus on both basic and applied aspects of the subject. Since bioremediation is an important tool in detoxifying and eliminating environmental contaminants, a thorough understanding of microbial genetics, biochemistry, and physiology is required. Attempts should be made to bridge the gap between success at laboratory level and success of the same at a field scale. Many times, laboratory testing does not accurately predict field results for many processes. The reason for the most part is attributed to differences in physiological conditions, concentration of the target chemical, and other physical, chemical, and microbial aspects that either was not taken into consideration or show constant variation. Research should focus on studies that are closer to "real" field or ground conditions. The concentration of the target chemicals used for carrying out biodegradation studies in the laboratory should not be hypothetical but should relate to contamination levels present in the environment. Further, treatment of hazardous chemicals in the environment also presents the possibility of unknown by-products of biodegradation entering the environment.

References

1. Ollgaad H, Frost L, Galster J, Hensen OC (1999) Survey of azocolorant on Denmark: Milgoproject 509. Danish Environmental Protection Agency
2. Stolz A (2001) Appl Microbiol Biotechnol 56:69–80
3. Ishikawa Y, Ester T, Leader A (2000) Chemical economics hand book: dyes. Svi. Chemical and Health Business Services, Menlo Park, CA
4. Will R, Ishikawa Y, Leader A (2000) Synthetic dyes. In: Chemical economics handbook, synthetic dyes. Svi. Chemical and Health Business Services, Menlo Park CA
5. O'Neill C, Hawkes FR, Hawkes DW, Lourenco ND, Pinheiro HM, Delee W (1999) J Chem Technol Biotechnol 74:1009–1018
6. Zollinger H (1987) Color chemistry-synthesis, properties and application of organic dyes and pigments. VCH, New York, pp 92–102
7. Gordon PF, Gregory P (1983) Organic chemistry in colour. Springer, New York, p 322
8. APHA (1998) Standard methods of water and wastewater 19th edition. American Public Health Association, Washington DC, USA
9. Banat IMP, Singh D, Marchant R (1996) Bioresour Technol 58:217–227
10. Delee W, O'Niell C, Hawkes FR, Pinheiro HM (1998) J Chem Technol Biotechnol 73:323–325
11. Hao OJ, Kim H, Chiang PC (2000) Crit Rev Environ Sci Technol 30:449
12. Kuhad RC, Good H, Teiputhi KK, Singh A, Ward OP (2005) Adv Appl Microbiol 56:185–213
13. McMullan G, Meehan C, Connedy A, Kirby H, Robinson T, Nigam P, Banat IM, Merchant R, Smyth WF (2001) Appl Microbiol Biotechnol 56:81–87
14. Pandey A, Singh P, Iyengar L (2007) Int Biodeterior Biodegradation 59:73–84
15. Pearce CI, Lloyd JR, Guthrie JT (2003) Dyes Pigm 58:179–196
16. Rai H, Bhattacharya M, Singh J, Bansuel TK, Vats P, Banergee UC (2005) Crit Rev Environ Sci Technol 35:219–238
17. Sandhya S, Swaminathan K, Swaminathan T (2007) Decolorization and treatment of recalcitrant dye industry wastewater containing azo dyes. In: Trivedi PC (ed) Industrial pollution and its management. Aavishkar, India, pp 148–171
18. Vander Zee FP, Villaverde S (2005) Water Res 39:1425–1440
19. Walker R, Ryan AJ (1971) Xenobiotica 1(4/5):483–486
20. Bras R, Gomes M, Ferra MIA, Pinheiro HM, Gonclaves IC (2005) J Biotechnol 115:57–66
21. Guo J, Zhou D, Wang C, Titan C, Wang P, Salah Udin M, Yu H (2007) Water Res 41:426–432
22. Mendez-Paz D, Omil F, Lema JM (2005) Water Res 39:771–778
23. Rajaguru P, Kalaiselvi K, Palanival M, Subburam V (2000) Appl Microbiol Biotechnol 54:268–273

24. Singh P, Saghi R, Pandey A, Iyengar L (2007) Bioresour Technol 98:2053–2056
25. Chinrelkitvanich S, Tuntoolvest M, Panswad T (2000) Water Res 43:2223–2232
26. Isik M, Sponza DT (2005) Process Biochem 40:1189–1193
27. Talarposhti AM, Donnelly T, Anderson GK (2001) Water Res 35:425–432
28. Bras R, Ferra IA, Pinheiro HM, Goncalves IC (2001) J Biotechnol 89:155–162
29. Plumb TT, Bell J, Stuckey DC (2001) Appl Environ Microbiol 67:3226–3235
30. Yoo ES, Libra J, Adrian L (2001) J Environ Eng (ASCE) 127:844–849
31. Sani RK, Banerjee UC (1999) Enzyme Microbial Technol 24:433–437
32. Nigam P, Banat IM, Singh D, Marchant R (1996) Proc Biochem 31:495–42
33. Hu TL (2001) Water Sci Technol 43(2):261–269
34. Hitz HR, Huber W, Reed RH (1978) JCDC 94:71–76
35. Kulla HG (1981) Aerobic bacterial degradation of azo dyes. In: Leisinger T, Cook AM, Nüesch J, Hütter R (eds) Microbial degradation of xenobiotics and recalcitrant compounds. Academic, London, pp 387–399
36. Beydilli MI, Pavlostathis SG, Tincher WC (2000) Water Environ Res 72(6):698–705
37. Zimmermann T, Gasser F, Kulla HG, Leisinger T (1984) Arch Microbiol 138:37–43
38. Zimmermann T, Kulla HG, Leisinger T (1982) Eur J Biochem 129:197–203
39. Vander Zee FP, Bisschops IAE, Blanchard VG, Bouwman RHM, Lettinga G, Field JA (2003) Water Res 37:3098–3109
40. Russ R, Rau J, Stoz A (2000) Appl Environ Microbiol 66:1429–1434
41. Moir D, Masson S, CXhu I (2001) Environ Toxicol Chem 20:479–484
42. Levine WG (1991) Drug Metab Rev 23:253–309
43. Dos Santos AB, Cervantes FJ, Yaya-Beas RE, van Lier JB (2003) Enzyme Microb Technol 33:942–951
44. Keck A, Klein J, Kundlich M, Stolz A, Knackmuss HJ, Mattes R (1997) Appl Environ Microbiol 63:3684–3690
45. Field JA, Brady J (2003) Water Sci Technol 48:187–193
46. Sandhya S, Padmavathy S, Swaminathan K, Subrahmanyam YV, Kaul SN (2005) Proc Biochem 40:885–890
47. Rafii F, Freankalin W, Cerniglia CE (1990) Appl Environ Microbiol 56:2146–2151
48. Rafii F, Cerniglia CE (1995) Environ Health Perspect 103:17–19
49. Rafii F, Colemann T (1999) J Basic Microbiol 39:29–35

Biodegradation of Azo Dyes in Anaerobic–Aerobic Sequencing Batch Reactors

Özer Çinar and Kevser Demiröz

Abstract Effluent discharge from textile and dyestuff industries to neighboring water bodies is currently causing significant health concerns to environmental regulatory agencies due to the toxicity, mutagenicity, and carcinogenicity of the dyes and their breakdown products. Therefore, considerable attention has been given to evaluate the removal of dyes during wastewater treatment and in the natural environment. The most widely used dyes in industries are azo dyes, which require an anaerobic and aerobic phases for their complete biodegradation. Anaerobic stage is the first step of the treatment process in which azo dyes are reduced, resulting in toxic and colorless aromatic amines. Since breakdown products of azo dyes, which are formed when the azo bond is cleaved and color is removed, are resistant to anaerobic biodegradation, aerobic phase is therefore essential for complete biodegradation of colored effluents. Biological treatment has long been known, and the use of sequencing batch reactors (SBRs) for treating textile wastewater has attracted interest. The cyclic operations of SBR provide both color removal in anaerobic stage and aromatic amine removal in aerobic stage.

Keywords Anaerobic–aerobic treatment, Aromatic amine removal, Azo dye, Decolorization, Sequencing batch reactor

Contents

Ö. Çinar (✉)
Department of Environmental Engineering, Faculty of Engineering and Architecture, Kahramanmaraş Sütçü Imam University, 46060 Kahramanmaraş, Turkey
e-mail: ocinar@alumni.clemson.edu

Kevser Demiröz
Department of Environmental Engineering, Suleyman Demirel University, Isparta, Turkey

H. Atacag Erkurt (ed.), *Biodegradation of Azo Dyes*,
Hdb Env Chem (2010) 9: 59–72, DOI 10.1007/698_2009_44,

Abbreviations

AQDS	Anthraquinone-2,6-disulfonate
AQS	Anthraquinone-2-sulfonate
FAD	Flavin adenide dinucleotide
FMN	Flavin adenide mononucleotide
HRT	Hydraulic retention time
NADH	Nicotinamide adenine dinucleotide
NAD(P)H	Nicotinamide adenine dinucleotide phosphate
SBR	Sequencing batch reactor
SRB	Sulfate reducing bacteria
SRT	Sludge retention time

1 Introduction

The control of water pollution has become of increasing importance in recent years due to the increase in population, development, and assortment of industries. In particular, increasing demand for textile products is making textile industry one of the main sources of water pollution problems. In fact, the main problem comes from dyes, which are mainly used to color synthetic and natural fabrics. Azo dyes are the largest class of dyes used due to more economical synthesis. Because the release of dyes together with their breakdown products into the environment constitutes serious problems, the release of colored effluents into the environment is undesirable [1]. Interestingly, although most dyes have a low toxicity, their components and breakdown products can be more toxic. To overcome this problem, much attention has been focused on the effective treatment of dyes discharged from the dying and textile industries.

Biological methods are commonly considered to be the most effective treatment applications since they present lower operating costs and improved applicability [2, 3]. Biological processes applied for decolorization of textile effluents are based on anaerobic and aerobic treatment. While anaerobic treatment provides reductive cleavage of the dyes' azo bond, aerobic treatment of azo dyes has been proven ineffective in most cases, but is often the typical method of treatment used today [4–8]. Though anaerobic treatment removes the color of the dye, aromatic amines resulting from decolorization process in the anaerobic treatment are not mineralized under anaerobic conditions and tend to accumulate to toxic levels [7, 9]. Single-step anaerobic processes also have limitations in terms of low chemical oxygen demand (COD) removal [10]. To remove the breakdown products of dyes from the wastewater, it has been frequently reported that aerobic conditions are essential in which many aromatic amines are readily further mineralized as well as remaining COD. Combination of anaerobic and aerobic conditions is therefore the most convenient concept for treating colored wastewaters [6, 8, 11–13]. So far, so many reactor types have been used to provide effective treatment for textile wastewaters, such as fluidized bed, upflow anaerobic sludge blanket reactors, and packed bed reactors. Recent studies have indicated the success of sequential biological systems in achieving the complete biodegradation of azo dyes. That fill, react, settle, and draw operations can be provided in a single reactor make the sequencing batch reactor (SBR) operation flexible. In SBR operations, a cycle is repeated continuously and all the operations can be achieved in a single reactor. SBR has the ability of achieving complete biodegradation of azo dyes by providing decolorization of textile wastewaters with the added possibility of metabolite mineralization in the aerobic period. SBR can tolerate often-variable organic loads, since reaction time can be adjusted to the feed load.

2 Anaerobic Color Removal

As mentioned in the previous section, anaerobic phase is the first stage of decolorization process starting with the formation of intermediary aromatic amines by reductive cleavage of the azo bond [4–6, 14]. The research papers reviewed suggests that color removal is mainly associated with the anaerobic stage of the SBR; however, contribution of aerobic stage is almost none. In fact, anaerobic decolorization process is based on oxidation reduction reactions in which azo dye acts as an electron acceptor. Therefore, to achieve effective decolorization, anaerobic conditions with a low redox potential (<-50 mV) is desired [15]. Under anaerobic conditions, azo dye, which is characterized with one or more –N=N– groups, is reduced and cleavage of azo bond is achieved. Decolorization process under these conditions requires an organic carbon/energy source. Glucose, starch, acetate, whey, and tapioca are the simple substrates used as organic carbon source in the SBR studies, which acts as electron donor in the anaerobic decolorization process [16, 17]. Electrons released from the oxidation of electron-donating primary substrate are transferred to the electron-accepting azo dye by carriers of the electron transport chain, thereby resulting in the color removal.

Anaerobic azo dye reduction can be mediated by enzymes, low molecular weight redox mediators, and chemical reduction by biogenic reductants. These reactions can be located either intracellular or extracellular. Reduction of highly polar azo dyes, which cannot pass through the cell membranes, is located outside the cell. Like azo dyes, nicotinamide adenine dinucleotide phosphate, which is believed to be the main source of electrons, also cannot pass through the cell membranes. Azo reductase enzyme, which is oxygen-sensitive and released extracellularly, is found to be responsible for the reduction of azo dyes.

Anaerobic stages of SBR studies have been shown efficient with color removal rates mostly higher than 70% [3, 18–20]. Meanwhile, COD removal efficiency of anaerobic phase of SBR was found to depend on dyestuff type, amount of initial COD concentration, anaerobic cycle time, etc. Nevertheless, there are also reports about no efficient COD removal in anaerobic cycle of SBR [21, 22].

3 Factors Effecting Anaerobic Color Removal Efficiency

This review article summarizes the results of several research studies dealing with combined anaerobic–aerobic SBRs. Since anaerobic stage is the first and the most important phase for color removal, parameters affecting color removal should be determined to operate SBRs efficiently. Therefore, this review especially presents the problems dealing with anaerobic phase of SBRs. Since most of the azo dyes can be decolorized under anaerobic conditions, anaerobic biodegradation seems to be nonspecific. Nevertheless, decolorization can be affected by so many parameters such as organic carbon source added, microorganisms selected, dye structure, cycle time, sludge age, and alternative electron acceptors involved. Therefore, factors affecting anaerobic color removal efficiency are briefly discussed in subsequent sections.

3.1 Microorganisms

In most of the reported processes of azo dye biodegradation, a wide range of organisms are found to reduce azo compounds, such as bacteria, algae, and fungi. Azo dyes are generally known to resist aerobic bacterial biodegradation with the exception of bacteria with specialized azo dye reducing enzymes. Bacterial strains that can aerobically reduce azo dyes cannot use dye as the growth substrate and therefore require organic carbon sources. There are only a few bacteria that are able to grow on azo dyes as the sole carbon source. Aromatic amines resulting from the reductive cleavage of azo bond can be used as a carbon and energy source for bacterial growth. Like carbon source, a nitrogen source is also essential for decolorization process, with the exception of bacteria that can use azo dyes as a nitrogen source. As reported earlier, ammonium chloride is the most suitable among all nitrogen sources for SBR studies, since it is believed that nitrate is a better electron acceptor than azo bond [23].

Based on the previous publications, azo dye can be reduced by azoreductase-catalyzed reduction under anaerobic conditions. But still there is a speculation whether bacterial flavin reductases are responsible for the azo reductase activity observed with bacterial cell extracts. In a published report, it is reported that flavin reductases are indeed able to act as azo reductases [24]. Bacteria produce extracellular oxidative enzymes, which are relatively nonspecific enzymes catalyzing the oxidation of a variety of dyes. It was reported that so many diverse groups of bacteria play a role in decolorization. It has been also reported that mixed microbial community could reduce various azo dyes, and members of the γ-proteabacteria and sulfate reducing bacteria (SRB) were found to be prominent members of mixed bacterial population by using molecular methods to determine the microbial population dynamics [1].

3.2 *Dye Structure*

It appears that almost every azo compound that has been tested is biologically reduced under anaerobic conditions; nevertheless, though similar conditions were provided, different color removal efficiencies were achieved. This indicates that dye structure is important when investigating biological color removal by SBRs. It was reported that metal-ion containing dyes can have adverse effect on decolorization efficiency [25, 26]. It was also reported that azo compounds with methyl, methoxy, sulpho, or nitro groups are being less likely to be biodegraded than the others with a hydroxyl or amino group [27, 28]. Azo dyes with a limited membrane permeability, such as sulfonated azo dyes, cannot be reduced intracellularly [29].

3.3 *Cycle Time*

Though cycle time plays an important role in the SBR for the decolorization process, not many reports are found in the literature. The long retention times are often applied in the anaerobic phase of the reactor studies, such as 18 and 21 h. In several studies, it was reported that there is a positive correlation between the anaerobic cycle time and the color removal [30, 31]. Indeed, in combined anaerobic–aerobic SBRs, since bacteria shifted from aerobic to anaerobic conditions, or vice versa, anaerobic azo reductase enzyme can be adversely affected by aerobic conditions, which is essential for aromatic amine removal, thereby resulting in insufficient color removal rate. To investigate the effect of cycle time on biodegradation of azo dyes, Çinar et al. [20] operated SBR in three different total cycle times (48-, 24- and 12-h), fed with a synthetic textile wastewater. The results indicated that with a decrease in anaerobic cycle time, the system performance on color removal is not adversely affected; on the contrary, both color removal efficiency and COD removal efficiency are slightly improved.

3.4 Sludge Age

The sludge retention time (SRT) is known as a very important operational parameter for color removal in SBR system. To obtain efficient color removal rate, adequate microbial population is desired. It was reported that 10 days SRT remained insufficient to obtain adequate population, and to ensure the color removal, SRT was increased to 15 day [2].

3.5 Alternative Electron Acceptors

Decolorization of azo dyes starts by reductive cleavage of azo bond. Electrons releasing from oxidation of organic compounds in the wastewaters goes through the azo dye and cleaves the azo bond. As anaerobic color removal occurs by the way of reduction of the azo dye, which acts a final electron acceptor in the microbial electron transport chain, existing different electron acceptors in anaerobic zone can be assessed as limiting factor for the dye removal. Alternative electron acceptors such as oxygen, nitrate, sulfate, and ferric ion may compete with the azo dye for reducing equivalents, resulting in insufficient color removals under anaerobic conditions.

3.5.1 Oxygen

Anaerobic reactors in full-scale treatment systems are designed as open to the atmosphere. The effect of oxygen entering anaerobic reactors through the surface is generally assumed to be negligible since surface area is small relative to the reactor volume. Oxygen can get into the anaerobic reactors of wastewater treatment plants, with the mixed liquor recirculated from the aerobic zone and mixing. The impact of oxygen on anaerobic color removal efficiency becomes progressively larger when it is thought that oxygen is the most effective electron acceptor on the electron transport chain.

Researchers have reported that decolorization is significantly affected from the high-redox-potential electron acceptors and dissolved oxygen. This is because electrons released by oxidation of organic compounds are preferentially used to reduce oxygen rather than the azo dye. Oxygen has an adverse effect on decolorization under anaerobic conditions; therefore, facultative or obligate anaerobes are necessary for azo dye reduction [32]. Inhibition of azo reductase activity by oxygen was also reported for *Pseudomonas luteola* [14, 33]. Indeed, nicotinamide adenine dinucleotide (NADH) leads to bacterial biodegradation of azo dyes by acting as electron donor. In spite of the fact that oxygen is an electron acceptor, the consumption of NADH by oxidative phosphorylation can adversely affect the enzymatic decolorization of azo dye. A recent study results also suggested that the presence of oxygen inhibits azo decolorization when the dissolved oxygen concentration in the medium was higher than 0.5 mg/L [34]. This is mainly due to the adverse effect of the molecular oxygen on anaerobic azo reductase enzyme.

3.5.2 Nitrate

Nitrate is normally found in textile processing wastewaters and generally comes from salts such as sodium nitrate, which is included in the dye baths for the improvement of dye fixation to the textile fibers. Nitrate concentrations used in textile processing can reach 40–100 g/L [35]. The importance of nitrate in anaerobic phase of SBR is that nitrate can compete with the azo dye for reducing the equivalents formed, resulting in decreasing decolorization [2, 5, 7, 35, 36]. Wuhrmann et al. [5] reported that azo dye cannot be decolorized until denitrification ends up.

3.5.3 Sulfate

Like nitrate, sulfate is also a constituent of textile processing wastewaters. Sulfate is generally added to the dye baths for ionic strength adjustment or it may be formed by the oxidation of sulfur species used in dyeing processes, such as sulfide, hydrosulfide, and dithionite [37].

There are so many reports highlighting different effects of sulfate on azo dye degradation. It seems that, in the presence of sulfate, decolorization may be rather stimulated than competitively suppressed [7, 31, 35–37]. It was reported that when sulfate-reducing activity of microbial population in SBR is inhibited by the addition of molybdate, anaerobic azo dye removal efficiency is decreased. Indeed, since sulfate acts as an electron acceptor under anaerobic conditions, it may compete with the dyes for the electrons available, thus causing an adverse effect on the decolorizing process. However, microbial population and sulfate concentration are also important for the reactions taking place during anaerobic phase. High sulfate concentrations are found to adversely affect decolorization unless sufficient amount of substrate is supplied to overcome the negative effects of elevated concentrations of sulfate [38]. Furthermore, when sulfate is reduced under these conditions by SRB, sulfide, which is known as bulk reductant, is generated and can in turn serve as an electron donor. Sulfide generation is found to contribute to the reduction of azo dyes. It is also reported that cofactors involved during microbial reduction of sulfate, such as cytochrome C3 (–205 mV) and NADH (–324 mV), have appropriate redox potential. Therefore, they can channel the electrons to azo dyes. Meanwhile, the redox potentials of the dye reduction that are more positive than the redox potential of biological sulfate reduction (–220 mV) can be accelerated by sulfate.

3.5.4 Ferric Iron

Ferric iron can act as an electron acceptor under the anaerobic conditions the azo dye is in. Like sulfate, it was found that addition of ferric iron to the reactor stimulates the azo dye reduction. Indeed, the reactions are dealing with the redox couple Fe (III)/Fe (II), which can act as an electron shuttle for transferring electrons from electron donor to the electron accepting azo dye. Meanwhile, reactions of both reduction of Fe (III) to Fe (II) and oxidation of Fe (II) to Fe (III) facilitate the electron transport from the substrate to azo dye, thus acting as an extracellular redox mediator [31].

3.5.5 Redox Mediators

Since long retention times are often applied in the anaerobic phase of the SBR, it can be concluded that reduction of many azo dyes is a relatively a slow process. Reactor studies indicate that, however, by using redox mediators, which are compounds that accelerate electron transfer from a primary electron donor (co-substrate) to a terminal electron acceptor (azo dye), azo dye reduction can be increased [39, 40]. By this way, higher decolorization rates can be achieved in SBRs operated with a low hydraulic retention time [41, 42]. Flavin enzyme cofactors, such as flavin adenide dinucleotide, flavin adenide mononucleotide, and riboflavin, as well as several quinone compounds, such as anthraquinone-2,6-disulfonate, anthraquinone-2,6-disulfonate, and lawsone, have been found as redox mediators [43–46].

Though accelerating effect of redox mediators is proved, differences in electrochemical factors between mediator and azo dye is a limiting factor for this application. It was reported that redox mediator applied for biological azo dye reduction must have redox potential between the half reactions of the azo dye and the primary electron donor [37]. The standard redox potentials for different azo dyes are screened generally between –430 and –180 mV [47].

3.6 Primary Electron Donor Type

Since anaerobic azo dye reduction is an oxidation–reduction reaction, a liable electron donor is essential to achieve effective color removal rates. It is known that most of the bond reductions occurred during active bacterial growth [48]. Therefore, anaerobic azo dye reduction is extremely depended on the type of primary electron donor. It was reported that ethanol, glucose, H_2/CO_2, and formate are effective electron donors; contrarily, acetate and other volatile fatty acids are normally known as poor electron donors [42, 49, 50]. So far, because of the substrate itself or the microorganisms involved, with some primary substrates better color removal rates have been obtained, but with others no effective decolorization have been observed [31]. Electron donor concentration is also important to achieve higher color removal rates. Since there are so many reactions involved in bioreactor, competition for reducing equivalents by other reactions may increase the required amount of primary substrate. Though in theory the amount of electron donor per millimole monoazo dye azo is 32 mg COD, it was reported in a study that, even if 60–300 times higher stoichiometric amount is used, more electron donor source is needed [51].

3.7 *Dye Concentration*

In several studies, large variations in dye concentrations have been applied in the reactor studies, and it was reported that dye concentration may play a role in the decolorization process. In the case of exceeding the reactor's biological azo dye reduction capacity, high dye concentration may adversely affect the dye removal efficiency and COD removal efficiency. Kapdan and Özturk [10] reported that increasing initial dyestuff concentration adversely affect the COD removal performance of SBR. Nevertheless, dye removal rate may be increased by increasing dye concentrations [52]. Some of the reactor studies have proved the possibility of azo dye toxicity to microorganisms involved in biodegradation. Though toxicity is related to dye concentration, dye type applied is also important [53] Metal-complex dyes and reactive dyes are known (from the literature) to have toxicity effect on decolorization process [54].

4 Degradation of Metabolites

Anaerobic azo dye reduction, the reductive cleavage of azo linkage, is the first stage in the complete anaerobic–aerobic degradation of azo dyes, resulting in aromatic amine accumulation. Aromatic amines, which are formed during anaerobic treatment, are generally colorless and hazardous; therefore, a convenient treatment is required. Though mineralization of the aromatic amines under aerobic conditions is more common, it was reported that a few aromatic amines that are characterized by the presence of hydroxyl and carboxyl groups can be mineralized under anaerobic conditions [55, 56]. As a result, combined anaerobic and aerobic conditions are essential for the complete biodegradation of colored wastewaters.

4.1 *Anaerobic Fate of Aromatic Amines*

In anaerobic environment, the aromatic compounds can mineralize through hydroxylation, carboxylation, and redox reactions due to enzymatic reactions. Many of the amines that are formed during the anaerobic decolorization of azo dyes are known as unstable under aerobic conditions. It is a challenge that aromatic amines are bias to autooxidation, yielding recalcitrant polymeric products [29]. That is why, so many researchers have focused on the degradation of such compounds by anaerobic microorganisms.

In anaerobic environment, the ring opening of the aromatic structures is carried out by hydroxylation and carboxylation reactions, and the breakdown products can be degraded in different pathways by enzymes, metabolism, and co-metabolism of the cells [57]. Many bacteria are capable of degrading aromatic compounds under both oxidizing and reducing conditions, and so have the ability to synthesize enzymes for both conditions. Unfortunately, limited studies report about the enzyme systems of microorganisms playing a role on azo dye degradation and aromatic amine mineralization under anaerobic conditions.

From the published reports, there are few aromatic amines found that can be mineralized under anaerobic conditions, such as naphthalene amines, which can be utilized as the sole organic carbon source by bacterial cultures [58]. Furthermore, it was reported that 2-aminonaphthyl sulfonate can be degraded or used as sulfur source by pure cultures [59, 60]. In many reports, however, it was found that sulfonated aromatic amines cannot be degraded under anaerobic conditions [61].

4.2 Aerobic Fate of Aromatic Amines

Aerobic biodegradation of aromatic amines formed during anaerobic stage seems more promising compared to anaerobic biodegradation of aromatic amines. Since aromatic amines can be easily biodegraded aerobically through hydroxylation and ring opening of the aromatic ring, it is suggested to combine the anaerobic cleavage of the azo dyes with the aerobic biodegradation of the accumulated amines [62–64]. However, it was reported that some aromatic amines are readily autoxidized in the presence of oxygen [20, 65].That is why researches focus on the determination of specialized cultures, which can mineralize aromatic amines under anaerobic conditions.

In aerobic degradation of aromatic compounds by microorganisms, catechol, protocatecuate, and gentisate play a key role, since they are ring cleavage substrates in which an exceptionally large number of peripheral pathways converge. These central intermediates are then cleaved by dioxygenases such as catechol 1,2 dioxygenase, catechol 2,3 dioxygenase, protocatecuate 3,4 dioxygesae, protocatecuate 4,5 dioxygenase, and gentisate 1,2 dioxygenase [66–68].

Previous studies suggested that enzymes responsible for the aromatic amine removal become more active when the color removal rates are high, which resulted in more aromatic amine production within the SBR. Beside this, adverse effect of anaerobic conditions on aerobic enzymes was also reported, and results indicated that their activities increased in aerobic stage and decreased in anaerobic stages due to the absence of dissolved oxygen [20, 69]. Viliesid and Lilly [70] found that the activity of catechol 1,2-dioxygenase is dependent on the dissolved oxygen concentration and is influenced by the oxygen concentration. It was also found that the activity of catechol 1,2-dioxygenase is likely to be low in systems with more limited oxygen concentrations.

5 Conclusions

Azo dye-containing wastewaters seems to be one of the most polluted wastewaters, which require efficient decolorization and subsequent aromatic amine metabolism. On the basis of the available literature, it can be concluded that anaerobic–aerobic SBR operations are quite convenient for the complete biodegradation of both azo dyes and their breakdown products. Nevertheless, like the other methods used for biological treatment, SBRs treating colored wastewaters have some limitations. Presence of forceful alternative electron acceptors such as nitrate and oxygen, availability of an electron donor, microorganisms, and cycle times of anaerobic and aerobic reaction phases can be evaluated as quite significant.

Though treatment of azo dye-containing wastewaters needs combined anaerobic–aerobic phases, microorganisms are subjected to continually alternating anaerobic and aerobic conditions. Thus, it is presumable that anaerobic enzymes involved in the azo dye reduction may be adversely affected by aerobic conditions, as well as aerobic enzymes involved in the aromatic amine mineralization may be adversely affected by anaerobic conditions. Since little is known about the regulations of the enzymes involved in complete biodegradation of colored wastewaters, this approach seems to need advanced investigation to improve color removal and aromatic amine mineralization.

References

1. Pandey A, Poonam S, Leela I (2007) Bacterial decolorization and degradation of azo dyes. Int Biodeterior Biodegradation 59:73–84
2. Lourenço ND, Novais JM, Pinheiro HM (2001) Effect of some operational parameters on textile dye biodegradation in a sequential batch reactor. J Biotechnol 89(2–3):163–174
3. Shaw CB, Carliell CM, Wheatley AD (2002) Anaerobic/aerobic treatment of coloured textile effluents using sequencing batch reactors. Water Res 36(8):1993–2001
4. Walker R (1970) The metabolism of azo compounds: a review of the literature. Food Cosmet Toxicol 8(6):659–676
5. Wuhrmann K, Mechsner K, Kappeler T (1980) Investigation on rate-determining factors in the microbial reduction of azo dyes. Appl Microbiol Biotechnol 9:325–338
6. Haug W, Schmidt A, Nortemann B et al (1991) Mineralization of the sulfonated azo dye mordant yellow 3 by a 6-aminonaphthalene-2-sulfonate-degrading bacterial consortium. Appl Microbiol Biotechnol 57:3144–3149
7. Carliell CM, Barclay SJ, Naidoo N et al (1995) Microbial decolourasition of a reactive azo dye under anaerobic conditions. Water SA 21:61–69
8. Kudlich M, Bishop P, Knackmuss H-J et al (1996) Synchronous anaerobic and aerobic degradation of the sulfonated azo dye Mordant Yellow 3 by immobilized cells from a naphthalenesulfonate-degrading mixed culture. Appl Microbiol Biotechnol 46:597–603

9. Gottlieb A, Shaw C, Smith A et al (2003) The toxicity of textile reactive azo dyes after hydrolysis and decolourisation. J Biotechnol 101(1):49–56
10. Kapdan IK, Özturk R (2005) Effect of parameters on color and COD removal performance of SBR: sludge age and initial dyestuff concentration. J Hazard Mater B 123:217–222
11. Zaoyan Y, Ke S, Guangliang S et al (1992) Anaerobic–aerobic treatment of a dye wastewater by combination of RBC with activated sludge. Water Sci Technol 26:2093–2096
12. Seshadri S, Bishop PL, Agha AM (1994) Anaerobic/aerobic treatment of selected azo dyes in wastewater. Waste Manage 14(2):127–137
13. Hu TL (1998) Degradation of azo dye RP2B by *Pseudomonas luteola*. Water Sci Technol 38:299–306
14. Blumel S, Contzen M, Lutz M et al (1998) Isolation of a bacterial strain with the ability to utilize the sulfonated azo compound 4-carboxy-4′- sulfoazobenzene as the sole source of carbon and energy. Appl Microbiol Biotechnol 64:2315–2317
15. Dos Santos AB, Cervantes FJ, van Lier JB (2007) Review paper on current technologies for decolourisation of textile wastewaters perspectives for anaerobic biotechnology. Bioresour Technol 98(12):2369–2385
16. van der Zee FP, Villaverde S (2005) Combined anaerobic–aerobic treatment of azo dyes – a short review of bioreactor studies. Water Res 39:1425–1440
17. Yoo ES, Libra J, Adrian L (2001) Mechanism of decolorization of azo dyes in an anaerobic mixed culture. J Environ Eng (ASCE) 127:844–849
18. Lourenço ND, Novais JM, Pinheiro HM (2000) Reactive textile dye colour removal in a sequencing batch reactor. Water Sci Technol 42:321–328
19. Kapdan IK, Alparslan S (2005) Application of anaerobic–aerobic sequential system to real textile wastewater for color and COD removal. Enzyme Microb Technol 36:273–279
20. Çinar Ö, Yaşar S, Kertmen M et al (2008) Effect of cycle time on biodegradation of azo dye in sequencing batch reactor. Process Saf Environ Protect 86:455–460
21. Lourenço ND, Novais JM, Pinheiro HM (2003) Analysis of secondary metabolite fate during anaerobic–aerobic azo dye biodegradation in a sequential batch reactor. Environ Technol 24 (6):679–686
22. Supaka N, Juntongjin K, Damronglerd S et al (2004) Microbial decolorization of reactive azo dyes in a sequential anaerobic–aerobic system. J Chem Eng 99:169–176
23. Wang X, Cheng X, Sun D et al (2008) Biodecolorization and partial mineralization of Reactive Black 5 by a strain Rhodopseudomonas palustris. J Environ Sci 20:1218–1225
24. Russ R, Rau J, Stolz A (2000) The function of cytoplasmic flavin reductases in the reduction of azo dyes by bacteria. Appl Environ Microbiol 66:1429–1434
25. Chung KT, Fulk GE, Egan M (1978) Reduction of azo dyes by intestinal anaerobes. Appl Microbiol Biotechnol 35:558–562
26. Brown MA, DeVito SC (1993) Predicting azo dye toxicity. Crit Rev Environ Sci Technol 23:249–324
27. Zimmermann T, Kulla H, Leisinger T (1982) Properties of purified orange II-azoreductase, the enzyme initiating azo dye degradation by Pseudomonas KF46. Eur J Biochem 129:197–203
28. Claus H, Faber G, Koenig H (2002) Redox-mediated decolorization of synthetic dyes by fungal laccases. Appl Microbiol Biotechnol 59:672–678
29. Stolz A (2001) Basic and applied aspects in the microbial degradation of azo dyes. Appl Microbiol Biotechnol 56:69–80
30. Kapdan IK, Tekol M, Sengul F (2003) Decolorization of simulated textile wastewater in an anaerobic–aerobic sequential treatment system. Process Biochem 38(7):1031–1037
31. Albuquerque MGE, Lopes AT, Serralheiro ML et al (2005) Biological sulphate reduction and redox mediator effects on azo dye decolourisation in anaerobic–aerobic sequencing batch reactors. Enzyme Microb Technol 36:790–799
32. Chang JS, Kuo TS (2000) Kinetics of bacterial decolorization of azo dye with Escherichia coli NO_3. Bioresour Technol 75:107–111

33. Chung KT, Stevens SEJ (1993) Degradation of azo dyes by environmental microorganisms and helminths. Environ Toxicol Chem 12:2121–2132
34. Xu M, Guo J, Sun G (2007) Biodegradation of textile azo dye by Shewanella decolorationis S12 under microaerophilic conditions. Appl Microbiol Biotechnol 76:719–726
35. Carliell CM, Barclay SJ, Shaw C et al (1998) The effect of salts used in textile dyeing on microbial decolourisation of a reactive azo dye. Environ Technol 19:1133–1137
36. Panswad T, Luangdilok W (2000) Decolorization of reactive dyes with different molecular structures under different environmental conditions. Water Res 34(17):4177–4184
37. van der Zee FP, Bisschops IAE, Lettings G et al (2003) Activated carbon as an electron acceptor and redox mediator during the anaerobic biotransformation of azo dyes. Environ Sci Technol 37:402–408
38. Cervantes FJ, Enrıquez JE, Petatan EG et al (2007) Biogenic sulphide plays a major role on the riboflavin-mediated decolourisation of azo dyes under sulphate-reducing conditions. Chemosphere 68:1082–1089
39. Kudlich M, Keck A, Klein J (1997) Localization of the enzyme system involves in anaerobic reduction of azo dyes by Sphingomonas sp. strain BN6 and effect of artificial redox mediators on the rate of azo dye reduction. Appl Microbiol Biotechnol 63:3691–3694
40. Keck A, Rau J, Reemtsma T et al (2002) Identification of quinoide redox mediators that are formed during thedegradation of naphthalene-2-sulfonate by Sphingomonas xenophaga BN6. Appl Microbiol Biotechnol 68:4341–4349
41. Cervantes FJ, van der Zee FP, Lettinga G (2001) Enhanced decolourisation of acid orange 7 in a continuous UASB reactor with quinones as redox mediators. Water Sci Technol 44:123–128
42. Dos Santos AB, Cervantes FJ, Yaya-Beas RE et al (2003) Effect of redox mediator AQDS on the decolourisation of a reactive azo dye containing triazine group in a thermophilic anaerobic EGSB reactor. Enzyme Microb Technol 33:942–951
43. Semde R, Pierre D, Geuskens G (1998) Study of some important factors involved in azo derivative reduction by *Clostridium perfringens*. Int J Pharm 161:45–54
44. Cervantes FJ, van Der Velde S, Lettinga G et al (2000) Competition between methanogenesis and quinone respiration for ecologically important substrates in anaerobic consortia. FEMS Microbiol Ecol 34:161–171
45. Rau J, Knackmuss HJ, Stolz A (2002) Effects of different quinoid redox mediators on the anaerobic reduction of azo dyes by bacteria. Environ Sci Technol 36:1497–1504
46. Rau J, Maris B, Kinget R et al (2002) Enhanced anaerobic degradation of polymeric azo compounds by Escherichia coli in the presence of low-molecular-weight redox mediators. J Pharm Pharmacol 54:1471–1479
47. Dubin P, Wright KL (1975) Reduction of azo food dyes in cultures of *Proteus vulgaris*. Xenobiotica 5:563–571
48. Nigam P, Banat IM, Singh D (1996) Microbial process for the decolorization of textile effluent containing azo diazo and reactive dyes. Process Biochem 31:435–442
49. Tan NCG, Prenafeta-Boldu FX, Opsteeg JL et al (1999) Biodegradation of azo dyes in cocultures of anaerobic granular sludge with aerobic aromatic amine degrading enrichment cultures. Appl Microbiol Biotechnol 51:865–871
50. Pearce CI, Christie R, Boothman C et al (2006) Reactive azo dye reduction by Shewanella Strain J18 143. Biotechnol Bioeng 95:692–703
51. O'Neill C, Lopez A, Esteves S et al (2000) Azo-dye degradation in an anaerobic–aerobic treatment system operating on simulated textile effluent. Appl Microbiol Biotechnol 53 (2):249–254
52. Cruz A, Buitron G (2001) Biodegradation of disperse blue 79 using sequenced anaerobic/aerobic biofilters. Water Sci Technol 44(4):159–166
53. Luangdilok W, Paswad T (2000) Effect of chemical structures of reactive dyes on color removal by an anaerobic–aerobic process. Water Sci Technol 42(3–4):377–382
54. Libra JA, Borchert M, Vigelahn L (2004) Two stage biological treatment of a diazo reactive textile dye and the fate of the dye metabolites. Chemosphere 56(2):167–180

55. Heider J, Fuchs G (1997) Anaerobic metabolism of aromatic compounds. Eur J Biochem 243:577–596
56. Kuhn EP, Suflita JM (1989) Anaerobic biodegradation of nitrogen-substituted and sulfonated benzene aquifer contaminants. Hazard Waste Hazard Mater 6(2):121–134
57. Griebler C, Safinowski M, Vieth A et al (2004) Combined application of stable carbon isotope analysis and specific metabolites determination for assessing in situ degradation of aromatic hydrocarbons in a tar oilcontaminated aquifer. Environ Sci Technol 38:617–631
58. Nachiyar CV, Rajkumar GS (2004) Mechanism of navitan fast blue S5R degradation by pseudomonas aeruginosa. Chemosphere 57:165–169
59. Wittch RM, Rast HG, Knackmuss HJ (1988) Degradation of naphthalene-2, 6 and naphthalene-1, 6-disulfonic acid by a Moraxella sp. Appl Environ Microbiol 54:1842–1847
60. Rozgaj R, Glancer SM (1992) Total degradation of 6-amnionaphthalene 2-sulfonic acid by a mixed culture consisting of different bacterial genera. FEMS Microbiol Ecol 86:229–236
61. Tan NCG, van Leeuwen A, van Voorthuizen EM et al (2005) Fate and biodegradability of sulfonated aromatic amines. Biodegradation 16:527–537
62. Kalyuzhnyi S, Sklyar V (2000) Biomineralisation of azo dyes and their breakdown products in anaerobic–aerobic hybrid and UASB reactors. Water Sci Technol 41(12):23–30
63. O'Neill C, Hawkes FR, Hawkes DW et al (2000) Anaerobic–aerobic biotreatment of simulated textile effluent containing varied ratios of starch and azo dye. Water Res 34 (8):2355–2361
64. Tan NCG (2001) Integrated and sequential anaerobic/aerobic biodegradation of azo dyes. PhD Thesis agrotechnology and food sciences sub-department of environmental technology. Wageningen University, Wageningen, The Netherlands
65. O'Neill C, Hawkes FR, Esteves SRR et al (1999) Anaerobic and aerobic treatment of a simulated textile effluent. J Chem Technol Biotechnol 74:993–999
66. Altenschmidt U, Oswald B, Steiner E et al (1993) New aerobic benzoate oxidation pathway via benzoyl-coenzyme A and a 3-hydroxybenzoyl-coenzyme A in a denitrifying Pseudomonas sp. J Bacteriol 175:4851–4858
67. Çinar Ö (2002) Factors influencing biodegradation of benzoate by denitrifying bacterial enrichment cultures. PhD Dissertation Clemson University, Clemson SC
68. Wang CL, You SL, Wang SL (2006) Purification and characterization of a novel catechol 1, 2-dioxygenase from *Pseudomonas aeruginosa* with benzoic acid as a carbon source. Process Biochem 41:1594–1601
69. Çinar Ö, Deniz T, Grady CPL Jr (2003) Effect of oxygen on the stability and inducibility of biodegradative capability for benzoate. Water Sci Technol 48:247–254
70. Viliesid F, Lilly MD (1992) Influence of dissolved oxygen tension on the synthesis of catechol 1, 2-dioxygenase by Pseudomonas putida. Enzyme Microb Technol 14:561–565

Decolorization of Azo Dyes by Immobilized Bacteria

Rashmi Khan and Uttam Chand Banerjee

Abstract Synthetic organic dyes are essential for satisfying the ever growing demand in terms of quality, variety, and speed of coloration of large number of substances. Because of the xenobiotic nature of dyes, they are toxicant to biological system and causes serious damage to environment. Ever-increasing concerns about color in the effluent lead to the worldwide efforts to build up effective procedure for color elimination. Biodegradation is gaining popularity to clean up hazardous waste because of the clear picture of the costs and the benefits of microbial degradation. Removal of dyes from waste water is reviewed with respect to biological decolorization. Promising techniques with reference to biological treatment of wastewater are immobilization of microorganisms on different supports. Immobilization increases the stabilities of the enzyme at high pH and tolerance to elevated temperatures and to make the enzyme less vulnerable to inhibitors. Generally the covalent bonds during immobilization enhance stabilities of enzymes due to the limitation of conformational changes.

Keywords Bacterial degradation, Biodegradation, Immobilization, Unsulfonated dyes

Contents

R. Khan and U.C. Banerjee (✉)
Department of Pharmaceutical Technology (Biotechnology), National Institute of Pharmaceutical Education and Research, Sector 67, Mohali, Punjab, India
e-mail: ucbanerjee@niper.res.in

H. Atacag Erkurt (ed.), *Biodegradation of Azo Dyes*,
Hdb Env Chem (2010) 9: 73–84, DOI 10.1007/698_2009_45,

Abbreviations

6A2NS	Aminonaphthalene-2-sulfonate
FBR	Fluidized bed reactor
HPLC	High pressure liquid chromatography
HRT	Hydraulic retention time
MY3	Mordant yellow 3
PVA	Polyvinyl alcohol
TLC	Thin layer chromatography

1 Introduction

Highly colored substances are broadly known as colorants. Colorants are subdivided into dyes, which are soluble in the medium in which they are applied, and pigments are insoluble in the application medium [1]. Dyes are defined as colored substances that when applied to fibers give them an everlasting color. There are two important conditions for a colored compound to act as dye; first the presence of chromophore and second the presence of auxochromes [2, 3]. Chromophore has a potential to absorb light in near ultra violet region to produce color, and auxochromes help dye to get attached to fibers through stable chemical bonds (acidic groups –OH, –COOH, $-SO_3H$ and by basic groups $-NH_2$, NHR, $-NR_2$). By and large dyestuffs are made from one or more of the compounds obtained by the distillation of coal tar. Textile industries are the largest consumers of dyes, and it is estimated that 15–20% of the dyes is lost during the dyeing process and is released as effluents [4]. The major classes of dyes have anthroquinoid, indigoid, and azo aromatic compounds. The chemical structure of azo dyes is comprised of a conjugated system of double bonds and aromatic rings. All these structures allow strong π–π^* transitions in the UV–visible (UV–Vis) area, with high extinction coefficients. The drawback of these dyes is that they are not easily degraded by aerobic bacteria, and with the action of anaerobic or microaerobic reductive bacteria, they can form toxic or mutagenic compounds such as aromatic amines [5–7]. There is a great environmental concern about the fate of these azo dyes, with special emphasis on reactive dyeing of cellulosic fibers, where large amounts of unbound dyes are discharged in the effluent [8].

2 Toxicity Caused by Azo Dyes

Synthetic organic dyes are essential in fulfilling the ever growing demand, in terms of quality, variety, and speed for coloration of massive number of substances. Because of the toxic nature, these materials present certain hazards and environmental problems. Toxicity of dyes varies with azo dye structure. The problem of azo dyes regarding human toxicity is associated with the type of intermediates used in their synthesis. After the reduction or cleavage of the azo bond, aromatic amines are formed; these are used as colorants and can be reduced by intestinal anaerobes [9], thus becoming a risk for human health. The presence of such compounds in industrial wastewater creates serious environmental problems [10]. Azo dyes constitute a major class of environmental pollutants, accounting for 60–70% of all dyes and pigments used. The discharge of azo dyes into the environment is a concern due to coloration of natural waters and their absorption and reflection of sunlight falling in the water bodies. This interferes with the growth of bacteria and plants, causing an annoyance to the ecology of the receiving water body due to the toxicity, mutagenicity, and carcinogenicity of the dyes and their biotransformation products. Therefore, substantial attention has been given to evaluate the fate of azo dyes during wastewater treatment and in the natural environment. It is well reported in literature [11, 12] that azo dyes require an anaerobic and an aerobic phase for their complete biodegradation. Many azo dyes are substituted with a sulfonic acid group; special attention is paid to the sulfonated azo dyes and their biodegradation products, the sulfonated aromatic amines. The first step in the biodegradation of azo dyes is reduction of azo dye, resulting in the formation of aromatic amines. The potential problem of azo dyes regarding human toxicity is associated with the type of intermediates used in their synthesis and appears only after the reduction and cleavage of the azo bond to give aromatic amines. These amines are more hydrophobic and, thus, may easily cross the cell membranes, consequently being more toxic than the original dyes. Several azo dyes used as colorants for food, drugs, and cosmetics can be reduced by cell suspension of predominant intestinal anaerobes [9]. Ingestion of certain azo dyes is a risk for human health. In this sense, 1-amino-2-naphthol, produced by the reduction of *Acid Orange 7*, has been reported to stimulate bladder tumors [13]. The azo dye Amaranth, which was widely used as a food colorant in many countries [14], has been shown to be carcinogenic for rats [15]. Azo dyes based on benzidine or 2-napthylamine is considered genotoxic, and during the past decades, most of the researches in the dye manufacturing industry were focused on the production of an alternative product for these compounds.

Sulfonated azo dyes are widely used in different industries [16]. Some structure of sulfonated and unsulfonated azo dyes is shown in Fig. 1. These water-soluble azo dyes will enter the environment generally with wastewater discharge. Also, these sulfonated and unsulfonated azo dyes have a negative aesthetic effect on the wastewater, and some of these compounds and biodegraded products are also toxic, carcinogenic, and mutagenic [17]. There exists clear evidence that sulfonated azo dyes show decreased or no mutagenic effect compared to unsulfonated azo dyes

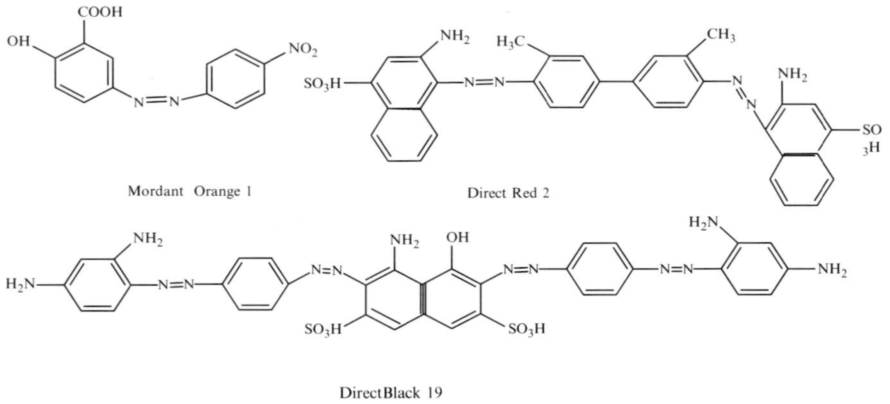

Fig. 1 Structure of sulfonated and unsulfonated azo dyes

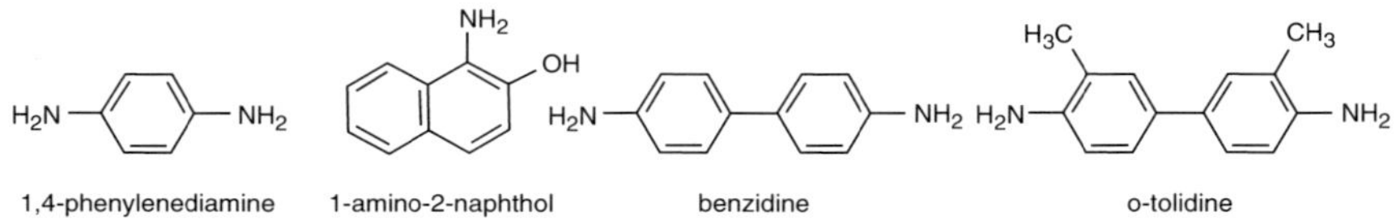

Fig. 2 Chemical structure of toxic degradation products

due to their electric charge and low lipophilicity, which prevents them from uptake and metabolic activation [5, 18, 19]. All the azo dyes containing a nitro group were found to be mutagenic [5], and a high toxicity of these azo dyes was also observed for methanogenic granular sludge [20]. Furthermore, some azo dyes can produce toxic degradation products. Examples of such harmful moieties are 1, 4-phenylenediamine, 1-amino-2-naphthol, benzidine, and substituted benzidines, such as *O*-tolidine, as shown in Fig. 2 [19, 21–23]. Because of the aforementioned effects, it is clear that azo dyes should not enter into the environment. An attractive method to prevent this is to apply biological treatment methods for their mineralization. Several other methods for azo dye removal from wastewater are also available, but they are mainly concerned to concentrate dyes.

3 Treatment of Dyes

Treatment of dye wastewater involves physical, physico-chemical, chemical, and biological methods. Physical processes are dilution, filtration, and gamma radiation. Physico-chemical includes adsorption, coagulation, flocculation, precipitation, reverse osmosis, ion exchange, etc.

3.1 Chemical Treatment of Azo Dyes

Chemical treatment includes oxidation, reduction, and adsorption by activated charcoal. Ozone treatment, precipitation, electrochemical treatment, and ion pair extraction are commonly used to treat the dye wastewater, but they produce a huge amount of sludge [24]. These methods are not only expensive but also generate wastes that are more difficult to dispose and less efficient with limited application [25]. The major techniques studied for the conventional color removals are activated charcoal, membrane technology, ozone treatment, and coagulation or flocculation methods. These techniques give significant results in color removal, volume capacity, operating speeds, and costs, though all of them have certain disadvantages associated with them. Activated charcoal method is extremely effective for color removal, but is capable of removing very small effluent at a very slow speed at a time and is very costly. Membrane technology, ozone treatment, and coagulation or flocculation are good for removal of large amount of effluent, but in case of membrane technology, it is fast and total operating cost is too high. Ozone treatment operates at moderate speed still very costly. Treatment processes such as sonocatalytic is a new interdisciplinary field, in which cavitations are induced by ultrasonic wave, which accelerate the course of chemical reaction. In early 1990s, people began to use cavitations caused by ultrasonic to degrade organic contaminants and disinfection. The technology has attracted attention for its convenience. However, because of its high operating cost, it has not been applied widely [26].

Azo dye treatment involves different mechanisms or locations such as enzymatic [27], non-enzymatic [28], intracellular [29], and extracellular [30]. These studies revealed that the azo dye, instead of being degraded by microorganisms, acts as an oxidizing agent for reduced flavin nucleotides of the electron transport chain. For example, ubiquitous sources of electrons, which is reduced forms of NAD(P)H, is able to reduce azo dyes in the absence of many enzymes [31]. Another extracellular reducing agent sulfide produced via respiration by sulfate-reducing bacteria also chemically decolorizes azo dyes [32, 33]. Azo dye reduction was greatly accelerated by the addition of redox mediators such as anthraquinone-sulfonate [34]. Human population is directly or indirectly exposed to dyes through their uses. So there is a need to search for a suitable technological application for the degradation of chromogenic dyes at large scale and also to find the new pathways of their conversion into beneficial by-products before discharging into aquatic or terrestrial ecosystem. An attractive method to prevent the aforementioned problems is to apply biological (microbial) treatment methods for their mineralization. Azo dyes are resistant to aerobic degradation; however, under anaerobic conditions, they can be reduced to potentially carcinogenic aromatic amines. Most of them are known environmental contaminants because of their complicated construction, different varieties, higher chemical stability, and poor biodegradation. In light of these facts, efficient removal of dyes from the environment has come to attention. However, as compared to the growth of the dye industries and the dye products, there is a little growth towards their removal methodologies. In this chapter, special emphasis is given on the immobilization of enzymes/cells, which is an emerging technique in the biological treatment of wastewater from dye and textile industries. Immobilization increases the stabilities of the enzyme even at higher pH, tolerance to elevated temperatures, and to make the enzyme less vulnerable to inhibitors.

3.2 Biological Treatment of Dyes

The biological treatment has many advantages over physical or chemical methods. There is a possibility of degradation of dye molecules to carbon dioxide and water and significantly less amount of sludge formation take place. Biological treatment, like aerobic or anaerobic process, is considered to be effective means of removing the bulk of pollutants from complex and high strength organic wastewater. Bacterial degradation of azo dyes is frequently initiated by an enzymatic biotransformation step that involves cleavage of azo linkages with the help of azo reductase via reduced coenzyme as the electron donor [35, 36]. It is well established that microorganisms play a crucial role in the mineralization of biopolymers and xenobiotic compounds [37]. It is reported that some azo dye biodegradation is enhanced under specific condition, particularly under nitrogen-limiting conditions [38], and its biodegradation depends on the chemical structure of the dye, nature of the substituents and their relative position [39, 40]. Aerobic and anaerobic treatment of dye wastewater helps in the reduction of azo dyes but are time consuming process, which is reflected by the requirement of long reaction time. Dye decolorization with immobilized whole cells is also an attractive procedure. Because of immobilization, biocatalysts show higher operational stability than free system. This is because immobilization usually enhances stabilities of enzymes, which are much less exposed to inhibitors such as halides, copper chelators, and dyeing additives than the free enzyme system [41]. The use of immobilized enzymes has significant advantages over soluble enzymes. In the near future, technology based on the enzymatic treatment of dyes present in the industrial effluent/wastewater will play a vital role. Treatment of wastewater on a large scale will also be possible by using reactors containing immobilized enzymes/whole cells.

4 Importance of Immobilization in Biological System

Immobilized cells have been extensively used for the production of useful and biologically important chemicals [42, 43], for the treatment of wastewaters [44–46], and for bioremediation of soil contaminated with numerous toxic chemicals. Immobilization not only simplifies separation and recovery of the immobilized bacteria and the binding agent, but it also makes the application reusable, which reduces the overall cost. Immobilized materials, furthermore, have comparatively longer operating lifetime due to an enhanced stability of the macromolecules or cells and, consequently, to protection from adverse conditions. Immobilized cell provides protection from higher concentration of recalcitrant organics that are toxic to free cells. Immobilized cells have been successfully employed as biocatalysts in environmental protection as well as in chemical, pharmaceutical, and food industry processes; there are very few reports of their direct application in the bioremediation of contaminated soils. Treating large amount of contaminated liquid or soil in bioreactors is technically challenging and costly.

5 Immobilized Bacteria

Immobilization of microorganism has been reported useful in biological wastewater treatment [47–50]. There are various methods of immobilization of bacterial cells. The overwhelming majority of the methods can be classified into four main categories: matrix entrapment, microencapsulation, adsorption, and covalent binding [51]. Among them, entrapment in polyvinyl alcohol (PVA) gel beads is the best, because of easy use, low cost, low toxicity to the system, and higher operational stability. Fang et al. in 2004 [52] reported that during decolorization some precautions should be taken to preserve mechanical strength of the immobilized beads to increase color removal. Cavities are formed inside the immobilized beads and make the beads soft and bulgy, which attributed to the disappearing of calcium alginate from the beads. In some literature it is also reported that only marginal turnover of the dyes was found under immobilized conditions with freely suspended cells [53]. Normally, enzyme immobilization is expected to provide stabilization effect restricting the protein unfolding process as a result of the introduction of random intra- and intermolecular cross links. Zille et al. in 2003 [54] reported less availability of the enzyme for interaction with anionic dyes due to the immobilization in a particular matrix. It was important to develop a wastewater clearing up technology for removing the dyes and the additional organic substances in a sole operation by dye assimilating bacteria [55–57]. Therefore, a rotating biological contactor with a disk on which *Pseudomonas cepacia* 13NA was immobilized using κ-carrageenan gel technique was developed. The results showed that dye-degradation activity was stable for a longer period of time.

On the basis of the microorganism's abilities to decolorize and the degradation of azo dyes and exploiting advantages of immobilized techniques, immobilization method was studied by different people. Some researchers used different types of stones and glasses as immobilization support for bacteria in aerobic or anaerobic condition [58, 59]. Dyes are eliminated by a wide variety of aerobic and anaerobic organisms, which are preferably employed as mixed cultures because of their relative toughness and versatility against xenobiotics compounds [60–62].

5.1 Immobilized Bacteria Under Aerobic Condition

Efforts to identify and isolate aerobic bacteria capable of degrading various dyes have been going on since more than two decades. Most dyes have long been considered nonbiodegradable or nontransformable under aerobic conditions [35]. Zeroual et al. (2001) [49] used bacterial strain *Enterobacter agglomerans* on different support materials by using fluidized bed reactor (FBR) and found that the bacteria had high effectiveness of decolorization of water polluted with azo dye methyl red. The immobilized cells of *E. agglomerans* exhibited higher capability for the complete decolorization of methyl red (100 mg/L) after 6 h of incubation under aerobic condition. Azo dyes are also reported to be decolorized in liquid solid fluidized bed reactor using PVA immobilized cell beads as support carriers [63]. The effect of cell bead number, density, initial dye concentration, hydraulic retention time, and diameter of immobilized cell beads on decolorization was studied in detail. It was concluded from the result that FBR with immobilized cell beads has 90% color removal efficiency with initial dye concentration $<$2,200 mg/L under continuous flow condition. To study the behavior of FBR on immobilized cell bead a model was designed. The model took into account both the mass transfer limitation and the hydrodynamic characteristics of immobilized cell beads in FBR.

Resmi et al. [59] used laterite stones for the immobilization of *Pseudomonas putida* (MTCC 1194). The amount of bacterial biomass attached to the support was 8.64 g/100 g of stones on dry weight basis. Packed bed reactor was used for treating mixture of seven azo dyes. With the help of immobilized bacterial strain, dye mixture was degraded to nontoxic smaller molecules. It was reported that even after 2 months, bacteria-coated pebbles were stable and suitable for the aerobic degradation of azo dyes. With the help of TLC and HPLC, 61.7% degradation was reported at the concentration of 50 μg/mL of dye.

5.2 Immobilized Bacteria Under Anaerobic Condition

Anaerobic decolorization of azo dyes was started back in the 1970s. Reticulated sintered glass was used as immobilization of anaerobic bacteria for the decolorization of wastewater and transformation of the azo dye to degradable products [58]. Full decolorization was achieved in less than 4 h HRT, and in addition to it, methane as biogas was also produced.

5.3 Immobilized Bacteria Under Aerobic–Anaerobic Condition

It is much clear from literature that most of the azo dyes are recalcitrant to aerobic degradation but can be degraded under anaerobic condition. Hence anaerobic treatment followed by aerobic treatment is mostly recommended for treating wastewater, specially decolorization of the wastewater from textile industries [1, 64, 65]. This condition can be implemented both by spatial separation of the anaerobic and aerobic waste via a sequential anaerobic–aerobic reactor system or inside one reactor, commonly termed as an integrated anaerobic–aerobic reactor system.

A collective anaerobic and aerobic treatment by immobilized microorganisms was first shown [66, 67] using calcium-alginate-immobilized co-cultures of a facultative anaerobic strain of *Enterobacter cloacae*. Reaction actually took place in the middle of the alginate beads. In these experiments, the reduced derivatives were oxidized in the outer parts of the alginate beads by a second aerobic strain (two different *Alcaligenes* species), which had the ability to oxidize 4-chloro-2-aminophenol.

The naphthalene sulfonate oxidizing bacterium *Sphingomonas* sp. BN6 was also reported for treating sulfonated azo dye mordant yellow 3 (MY3) under aerobic and anaerobic condition [53]. Under aerobic condition, degradation of dye was marginal. Under anaerobic condition, suspended cells of *Sphingomonas* cleaved the azo bond of MY3 to 6A2NS and 5-aminosalicylate. Immobilized cells under aerobic condition resulted in the formation of more than equimolar amounts of 5-aminosalicylate, but no 6A2NS. *Sphingomonas* sp. BN6 aerobically oxidizes 6A2NS to 5-aminosalicylate. It is hence concluded that cells in anaerobic center of the alginate beads reduced MY3 to 6A2NS and 5-aminosalicylate, and 6A2NS was oxidized to 5-aminosalicylate by cells immobilized in the outer aerobic zones of the alginate beads.

6 Future Work

For effective biological treatment of dye wastewater, immobilization of bacteria under aerobic anaerobic high-rate reactors should be given special attention. The main cause of effective treatment of these xenobiotics under immobilized condition in high rate reactors is the rapid facile reduction of these compounds to products of lower toxicity [68, 69]. Moreover, the immobilization of anaerobic bacteria and maintenance of a high concentration of biomass in the high rate reactors are factors that improve the tolerance of the anaerobic system to toxic substances [70, 71].

References

1. Rai HS, Bhattacharyya MS, Singh J et al (2005) Removal of dyes from the effluent of textile and dyestuff manufacturing industry: a review of emerging techniques with reference to biological treatment. Crit Rev Env Sci Technol 35:219–238
2. Austin GT (1994) Shreve's chemical process industries. McGraw Hill, New York
3. SBP Board of Consultants and Engineers (1994) Handbook of exported oriented dyes and intermediate industries. SBP Consultants and Engineers Pvt Ltd, India
4. Vaidya AA, Datye KV (1982) Environmental pollution during chemical processing of synthetic fibers. Colourage 14:3–10

5. Chung KT, Cerniglia CE (1992) Mutagenicity of azo dyes: structure activity relationship. Mutat Res 277:201–220
6. Chung KT, Stevens SE Jr (1993) Decolorization of azo dyes by environmental microorganisms and helminthes. Environ Toxicol Chem 12:2121–2132
7. Wong PK, Yuen PY (1996) Decolorization and biodegradation of methyl red by *Klebsiella pneumoniae* RS-13. Water Res 30:1736–1744
8. Pierce J (1994) Color in textile effluents the origins of the problem. J Soc Dyers Color 110:131–134
9. Chung KT, Fulk GE, Egan M (1978) Reduction of azo dyes by intestinal anaerobes. Appl Environ Microbiol 35:558
10. Hildenbrand S, Schmahl FW, Wodarz R et al (1999) Azo dyes and carcinogenic aromatic amines in cell culture. Int Arch Occup Environ Health 72:M52
11. Seshardi S, Bishop PL, Agha AM (1994) Anaerobic/aerobic treatment of selected azo dyes in wastewater. Waste Manage 14:127–137
12. Flores ER, Luijten M, Donlon BA et al (1997) Complete biodegradation of the azo dye azo disalicylate under anaerobic conditions. Environ Sci Technol 31:2098–2103
13. Bonser GM, Bradshaw L, Clayson DB et al (1956) A further study on the carcinogenic properties of ortho-hydroxyamines and related compounds by bladder implantation in the mouse. Br J Cancer 10:539–546
14. Collins TFX, McLaughlin J, Gray GC (1972) Teratology studies on food colorings. Part 1: embryo toxicity of Amaranth (FD & C Red no. 2) in rats. Food Cosmet Toxicol 10:619–624
15. Andrianova MM (1970) Carcinogenic properties of the red food dyes, amaranth SX Purple and 4R Purple. Vopr Pitaniya 29:61–66
16. Zollinger H (1987) Color chemistry – syntheses properties and applications of organic dyes pigments. VCH, New York
17. Grover IS, Kaur A, Mahajan RK (1996) Mutagenicity of some dye effluents. Nat Acad Sci Lett India 19(7–8):149–158
18. Levine WG (1991) Metabolism of azo dyes: implications for detoxification and activation drug. Metabol Rev 23(3–4):253–309
19. Rosenkranz HS, Klopman G (1990) Structural basis of the mutagenicity of 1-amino-2-naphthol-based azo dyes. Mutagenesis 5(2):137–146
20. Donlon BA, Razo-Flores E, Luijten M et al (1997) Detoxification and partial mineralization of the azo dye mordant orange 1 in a continuous upflow anaerobic sludge blanket reactor. Appl Microbiol Biotechnol 47:83
21. Chung KT, Fluk GE, Andrews AE (1981) Mutagenicity testing of some commonly used dyes. Appl Environ Microbiol 42(4):641–648
22. Reid TM, Morton KC, Wang CY et al (1984) Mutangencity of azo dyes following metabolism by different reductive/oxidative systems. Environ Mutagen 6:705–717
23. Rosenkranz HS, Klopman G (1989) Structural basis of the mutagenicity of phenylazoaniline dyes. Mutat Res 221(3):217–234
24. Cooper P (1995) Removing color from dye house wastewater. Asian Textile J 3:52–56
25. Chen KC, Huang WT, Wu JY et al (1999) Microbial decolorization of azo dyes by *Proteus mirabilis*. J Ind Microbiol Biotech 23:686–690
26. Liu Y (1995) Advancement of sono chemically degrade organic pollutants in waters. Adv Environ Sci (in Chinese) 3(4):77
27. Haug W, Schmidt A, Nortermann B et al (1991) Mineralization of the sulfonated azo dye mordant yellow3 by a 6-aminonaphthalene-2-sulfonate-degrading bacterial consortium. Appl Environ Microbiol 57:3144
28. Gingell R, Walker R (1971) Mechanisms of azo reduction by *Streptococcus faecalis* II The role of soluble flavins. Xenobiotica 1:231
29. Wuhrmann K, Menscher K, Kappeler T (1980) Investigation on rate determining factors in the microbial reduction of azo dyes. Eur J Appl Microbiol Biotechnol 9:325

30. Carliell CM, Barclay SJ, Naidoo N et al (1995) Microbial decolorization of a reactive azo dye under anaerobic conditions. Water 21:61
31. Nam S, Renganathan V (2000) Non-enzymatic reduction of azo dyes by NADH. Chemosphere 40:351–357
32. Yemashova N, Telegina A, Kotova I et al (2004) Decolorization and partial degradation of selected azo dyes by methanogenic sludge. Appl Biochem Biotechnol 119:31–40
33. Yoo ES (2002) Kinetics of chemical decolorization of the azo dye C.I Reactive Orange 96 by sulfide. Chemosphere 47:925–931
34. dos Santos AB, Bisschops IAE, Cervantes FJ et al (2004) Effect of different redox mediators during thermophilic azo dye reduction by anaerobic granular sludge and comparative study between mesophilic (30°C) and thermophilic (55°C) treatment for decolorization of textile wastewaters. Chemosphere 55:1149–1157
35. Zimmermann T, Kulla H, Leisinger T (1982) Properties of purified Orange II azo reductase the enzyme initiating azo dye degradation by *Pseudomonas* KF46. Eur J Biochem 129:197–203
36. Moutaouakkil A, Zeroual Y, Dzairi FZ et al (2003) Purification and partial chareterization of azoreductase from Enterobacter agglomerans. Arch Biochem Biophys 413:139–146
37. Loidl M, Hinteregger C, Ditzelmuller G et al (1990) Degradation of aniline and mono chlorinated anilines by siol born *Pseudomonas acidovorans* strains. Arch Microbiol 155:56
38. Spadaro JT, Gold MH, Renganathan V (1992) Degradation of azo dyes by the lignin-degrading fungus *Phanerochaete chrysosporium*. Appl Environ Microbiol 58:2397–2401
39. Pasti-Grigsby MB, Paszczynski A, Goszczynski S et al (1992) Influence of aromatic substitution patterns on azo dye degradability by *Streptomyces* spp. and *Phanerochaete chrysosporium*. Appl Environ Microbiol 58:3605–3613
40. Paszczynski A, Pasti-Grigsby MB, Goszczynski S et al (1992) Mineralization of sulfonated azo dyes and sulfanilic acid by *Phanerochaete chrysosporium* and *Streptomyces chromofuscus*. Appl Environ Microbiol 58:3598–3604
41. Rogalski J, Jozwik E, Hatakka A et al (1995) Immobilization of laccase from *Phlebia radiata* on controlled porosity glass. J Mol Catal A Enzym 95:99–108
42. Ohta T, Ogbonna JC, Tanaka H et al (1994) Development of a fermentation method using immobilized cells under unsterile conditions ethanol and L-lactic acid production without heat and filter sterilization. Appl Microbiol Biotechnol 42:246–260
43. Chang YC, Chou CC (2002) Growth and production of cholesterol oxidase by alginate-immobilized cells of Rhodococcus equi No 23. Biotechnol Appl Biochem 35:69–74
44. Gardea-Torresdey JL, Arenas NMC, Francisco KJ et al (1998) Ability of immobilized cyanobacteria to remove metal ions from solution and demonstration of the presence of metallothionein genes in various strains. J Hazard Subst Res 1:1–8
45. Chen KC, Chen JJ, Houng JY (2000) Improvement of nitrogen removal efficiency using immobilized microorganisms with oxidation-reduction potential monitoring. J Ind Microbiol Biotechnol 25:229–234
46. Wang CC, Lee CM, Lu CJ et al (2000) Biodegradation of 2, 4, 6-trichlorophenol in the presence of primary substrate by immobilized pure culture bacteria. Chemosphere 41:1873–1879
47. Chibata I, Tosa T (1981) Use of immobilized cells. Annu Rev Biophys Bioeng 10:197–216
48. Hyde FW, Hunt GR, Errede LA (1991) Immobilization of bacteria and *Saccharomyces cerevisiae* in poly (tetrafluoroethylene) membranes. Appl Environ Microbiol 57:219–222
49. Zeroual Y, Moutaouakkil A, Blaghen M (2001) Volatilization of mercury by immobilized bacteria (Klebsiella pneumoniae) in different support by using fluidized bed bioreactor. Curr Microbiol 43:322–327
50. Isaka K, Yoshie S, Sumino T et al (2007) Nitrification of landfill leachate using immobilized nitifying bacteria at low temperatures. Biochem Eng J 37:49–55
51. Monsan P (1982) Les methodes immobilisation enzymes. In: Durand G, Monsan P (eds) Les enzymes, productions utilizations industrielles. Gauthier-Villards, Paris, pp 81–118

52. Fang H, Wenrong H, Yuezhong L (2004) Investigation of isolation and immobilization of a microbial consortium for decoloring of azo dye 4BS. Water Res 38:3596–3604
53. Kudlich M, Bishop PL, Knackmuss HJ et al (1996) Simultaneous anaerobic and aerobic degradation of the sulfonated azo dye mordant yellow 3 by immobilized cells from naphthalenesulfonate degrading mixed culture. Appl Microbiol Biotechnol 46:597–603
54. Zille A, Tzanov T, Gubitz GM et al (2003) Immobilized laccase for decolorization of reactive black 5 dyeing effluent. Biotechnol Lett 25:1473–1477
55. Idaka E, Ogawa T, Horitzu H et al (1979) Isolation and identification of an azo assimilating bacterium II *Pseudomonas pseudomallei* 13NA. Res Rept Fac Eng Gifu Univ 29:68–70
56. Idaka E, Ogawa T, Sakaguchi M et al (1980) Characteristics of *Bacillus subtilis* azoreductase. Res Rept Fac Eng Gifu Univ 30:53–58
57. Idaka E, Ogawa T, Sakaguchi M et al (1979) Isolation and identification of an azo dye assimilating bacterium I *Aeromonas hydrophila* var 24B. Res Rept Fac Eng Gifu Univ 29:65–67
58. Georgiou D, Hatiras J, Aivasidis A (2005) Microbial immobilization in a two stage fixed bed reactor pilot plant for onsite anaerobic decolorization of textile wastewater. Enzyme Microb Technol 37:597–605
59. Resmi CS, Shaffiqu TS, Roy JJ et al (2008) Aerobic degradation of a mixture of azo dyes in a packed bed reactor having bacteria coated laterite pebbles. Biotechnol Prog 19:647–651
60. Binkley J, Kandelbauer A (2003) Effluent treatment-enzymes in activated sludge. In: Cavaco-Paulo A, Guebitz GM (eds) Textile processing with enzymes. Woodhead, Cambridge UK, pp 199–221
61. Keharia H, Madamvar D (2003) Bioremediation concept for treatment of dye containing wastewater: a review. Indian J Exp Biol 41:1068–1075
62. Sharma DK, Saini HS, Singh M et al (2004) Biodegradation of acid blue-15 a textile dye by an upflow immobilized cell bioreactor. J Ind Microbiol Biotechnol 31:109–114
63. Wua JY, Hwang SCJ, Chen CT et al (2005) Decolorization of azo dye in a FBR reactor using immobilized bacteria. Enzyme Microb Technol 37:102–112
64. Azmi W, Banerjee UC (2002) Biological stabilization of textile and dye stuff industrial waste. Indian Chem Eng Sec A 44:230–234
65. Zitomer DH, Speece RH (1993) Sequential environments for enhanced biotransformation of aqueous contaminants. Environ Sci Technol 27:227
66. Beunink J, Rehm HJ (1988) Synchronous anaerobic and aerobic degradation of DDT by an immobilized mixed culture system. Appl Microbiol Biotechnol 29:72–80
67. Beunink J, Rehm HJ (1990) Coupled reductive and oxidative degradation of 4-chloro-2-nitrophenol by a co-immobilized mixed culture system. Appl Microbiol Biotechnol 34:108–115
68. Donlon BA, Razo-Flores E, Field J et al (1995) Toxicity of N Substituted aromatics to acetoclastic methanogenic activity in granular sludge. Appl Environ Microbiol 61:3889
69. Soares GMB, Costa-Ferreira M, Pessoa de Amorim MT (2001) Decolorization of an anthraquinone type dye using a laccase formulation. Bioresour Technol 79:171
70. Dwyer DF, Krumme ML, Boyd SA et al (1986) Kinetics of phenol biodegradation by an immobilized methanogenic consortium. Appl Environ Microbiol 52:345
71. Pagga U, Brown D (1986) The degradation of dyestuffs: Part II. Behavior of dye stuffs in aerobic biodegradation tests. Chemosphere 15:479

Decolorization and Degradation of Azo Dyes by Redox Mediator System with Bacteria

Jianbo Guo, Li Kang, Xiaolei Wang, and Jingliang Yang

Abstract Azo dyes are the largest and the most diverse group of synthetic dyes widely used in many industries, which are generally recalcitrant to biodegradation due to their xenobiotic nature. The effective treatment of azo dye wastewaters has been a big challenge, and up to now there is no single and economically attractive treatment that can effectively decolorize dyes. However, notable achievements have been conducted to explore the accelerating effects of different redox mediators during the anaerobic decolorization and degradation of azo dyes over the last two decades. The accumulated evidence suggest that redox mediators play a major role of electron shuttles in the reductive decolorization of azo dyes, both by chemical and biological mechanisms. This review is focused on the bacterial decolorization and degradation of azo dyes catalyzed by redox mediators and the further investigation to enhance the applicability of redox mediators on the bio-transformation of azo dyes.

Keywords Azo dye, Bacteria, Decolorization, Degradation, Redox mediator

Contents

J. Guo (✉), L. Kang, X. Wang, and J. Yang
School of Environmental Science and Engineering, Hebei University of Science and Technology, 70 Yuhua East Road, Shijiazhuang 050018, P.R. China
e-mail: jianbguo@yahoo.com.cn

H. Atacag Erkurt (ed.), *Biodegradation of Azo Dyes*,
Hdb Env Chem (2010) 9: 85–100, DOI 10.1007/698_2009_46,

Abbreviations

9,10-AQS	Anthraquinone-2-sulfonate
AGS	Anaerobic granular sludge
AN	1-Amino-2-naphthol
AQ	9,10-Anthraquinone
AQDS	9,10-Anthraquinone-2,6-disulfonate
BV	Benzyl viologen
CBN12	5′deoxyadenosylcobalamin (vitamin B12)
CE	Enzyme cofactor
E	Bacteria/enzyme
ED	Primary electron donor
FAD	Flavin adenine dinucleotide
HEM	Hematine
HG	Hemoglobin
JUG	Juglone
ox	Oxidized
RA	Resazurin
Red	Reduced
RF	Riboflavin
RM	Redox mediator

1 Introduction

Azo dyes, which are aromatic compounds with one or more –N=N– groups, represent the largest and the most diverse group of synthetic dyes applied in a number of industries such as textile, food, cosmetics, and paper printing. All dyes do not bind to the fabric depending on the class of the dye. Its loss in wastewaters could vary from 2% for basic dyes to as high as 50% for reactive dyes, resulting in the release of dye-containing wastewater [1, 2]. Dye wastewaters are characterized by extreme fluctuations in many parameters such as chemical oxygen demand (COD), biochemical oxygen demand (BOD), pH, color, and salinity. The wastewater composition will depend on the different organic-based compounds, chemicals, and dyes used in the industrial dry and wet-processing steps [3, 4]. Willmott has reported that up to 1.56 mg dm^{-3} dye can be detected in receiving watercourses, although dye concentrations as low as 0.005 mg dm^{-3} are visible in clear river water [5, 6]. Therefore, the release of colored wastewaters into the environment is a serious environmental problem and a public health concern, not only because of their color, but also because many dyes from wastewater and their breakdown products are generally recalcitrant to biodegradation due to their xenobiotic nature [7].

The different technologies of dye removal, such as adsorption on inorganic or organic matrices, decolorization by photocatalysis or by oxidation processes, microbiological or enzymatic decomposition, have been developed by many researchers for the effective treatment of dyes from waters and wastewaters to decrease their impact on the environment [8–10]. The efficacy of these different technologies of dye removal was compared in many papers. And every technique has its technical and economical limitations. Generally, most physicochemical dye removal technologies have drawbacks because they are expensive, and greatly interfered by other wastewater constituents or generate waste products that must be handled. Alternatively, biological treatment may present a relatively inexpensive way to remove dyes from wastewater.

The bio-treatment technology of dyes, especially anaerobic azo dye reduction, has been thoroughly investigated, and most researchers agree that it is a nonspecific and presumably extracellular process in which reducing equivalents from either biological or chemical source are transferred to the dye.

During the last two decades, more studies have been conducted to explore the catalytic effects of different redox mediators on the bio-transformation processes. Redox mediators, also referred to as electron shuttles, have been shown to play an important role not only as final electron acceptor for many recalcitrant organic compounds, but also facilitating electron transfer from an electron donor to an electron acceptor, for example, azo dyes [8, 11, 12]. Redox mediators accelerate reactions by lowering the activation energy of the total reaction, and are organic molecules that can reversibly be oxidized and reduced, thereby conferring the capacity to serve as an electron carrier in multiple redox reactions.

This review article summarizes the bacterial decolorization and degradation of azo dyes catalyzed by redox mediators and the further investigation to enhance the applicability of redox mediators on the bio-transformation of azo dyes.

2 Mechanism of Azo Dye Decolorization and Degradation

Generally, the process of bacterial azo dye biodegradation consists of two stages. The first stage involves reductive cleavage of the dyes' azo bond, resulting in the formation of aromatic amines, which is generally colorless but potentially hazardous. The second stage involves degradation of the aromatic amines under aerobic conditions.

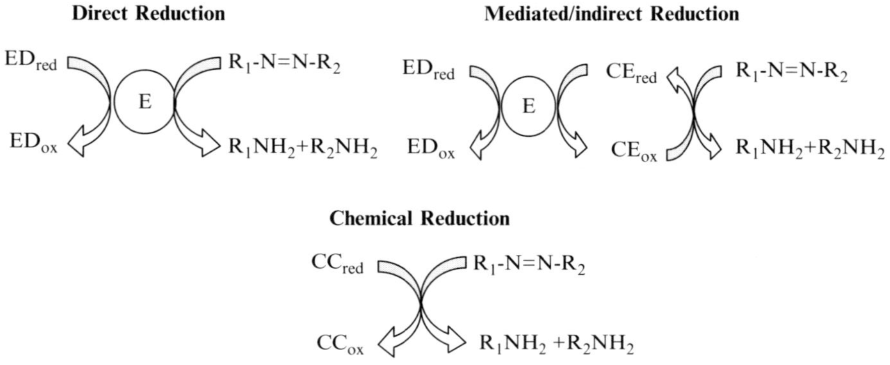

Fig. 1 Different bio-reduction mechanisms of azo dyes

Anaerobic bio-reduction of azo dye is a nonspecific and presumably extracellular process and comprises of three different mechanisms by researchers (Fig. 1), including the direct enzymatic reduction, indirect/mediated reduction, and chemical reduction. A direct enzymatic reaction or a mediated/indirect reaction is catalyzed by biologically regenerated enzyme cofactors or other electron carriers. Moreover, azo dye chemical reduction can result from purely chemical reactions with biogenic bulk reductants like sulfide. These azo dye reduction mechanisms have been shown to be greatly accelerated by the addition of many redox-mediating compounds, such as anthraquinone-sulfonate (AQS) and anthraquinone-disulfonate (AQDS) [13–15].

2.1 *Direct Enzymatic Azo Dye Reduction*

2.1.1 Aerobes

Under aerobic conditions, aerobic bacteria has so far been only found in studies capable of reducing azo compounds and produce aromatic amines by specific oxygen-catalyzed enzymes called azo reductases. These aerobic bacteria could grow with mostly simple azo compounds as sole source of carbon and energy and under strict aerobic conditions by using a metabolism that started with reductive cleavage of the azo linkage.

The azo reductases in aerobic bacteria were found to be existent when azoreductases from obligate aerobic bacteria were isolated and characterized from strains K22 and KF46 and were shown to be flavin-free after purification, characterization, and comparison 364, 362, 363. These intracellular azoreductases showed high specificity to dye structures. Furthermore, Blumel and Stolz cloned and characterized the genetic code of the aerobic azo reductase from *Pagmentiphaga kullae* K24 [16]. The gene encoded a protein with a molecular weight of 20,557 Da, having conserved a putative NAD(P)H-binding site in the amino-terminal region. Apart from these specific azoreductases, nonspecific enzymes catalyzing azo dye reduction also have been isolated from aerobically grown cultures of *Shigella dysenteriae*, *Escherichia coli*, and *Bacillus* sp. When characterized, these enzymes were found to be flavoproteins [15].

For these aerobic bacteria, the mono- and di-oxygenase enzymes are generally important to catalyze the incorporation of oxygen from O_2 into the aromatic ring of organic compounds prior to ring fission [17].

2.1.2 Strictly Anaerobes or Facultative Microorganisms

Under anaerobic or anoxic conditions, the azo bond (–N=N–) cleavage is conducted by specific enzymes (catalyzing only the reduction of azo dyes) or nonspecific enzymes (nonspecific enzymes that catalyze the reduction of a wide range of compounds, including azo dyes). However, there is no clear evidence for the specific azoreductases in anaerobically grown bacteria. Many anaerobic and facultative anaerobic strains may be responsible for the almost ubiquitous capacity of reducing azo dyes. Further research with the purified responsible enzyme from one of the strains showed that it was a flavoprotein capable of catalyzing the reduction of azo dyes as well as nitroaromatics [15]. Therefore, enzymatic anaerobic azo dye reduction is more or less a fortuitous reaction, where dye might act as an acceptor of electrons supplied by carriers of the electron transport chain. Alternatively, decolorization might be attributed to nonspecific extracellular reactions occurring between reduced compounds generated by the anaerobic biomass [18].

The azo bond (–N=N–) cleavage proceeds through two stages, involving a transfer of four-electrons (reducing equivalents). In each stage two electrons are transferred to the azo dye, which acts as a final electron acceptor (Fig. 2). The rate of decolorization is dependent on the added organic carbon source, as well as the dye structure. But there is no correlation between decolorization rate and molecular weight, indicating that decolorization is not a specific process and cell permeability is not important for decolorization.

At the same time, this mechanism of azo dye anaerobic reduction occurring intracellularly or extracellularly is still an argumentative question. Not all types of azo dyes could be reduced by intracellular azo dye reduction. For example, high molecular weight polymeric azo dyes and highly polar sulfonated azo dyes are unlikely to pass through the cell membrane, but the decolorization rates of sulfonated azo dyes increased by cell free-extracts, as well as by addition of toluene, that is, a membrane-active compound that increases cell lysis, thus showing the limited membrane permeability of this type of dye [53]. And the azo reductase was found to be located throughout the bacterial cytoplasm without showing association to membranes or other organized structures, but it was secreted before acting as an azoreductase in vivo.

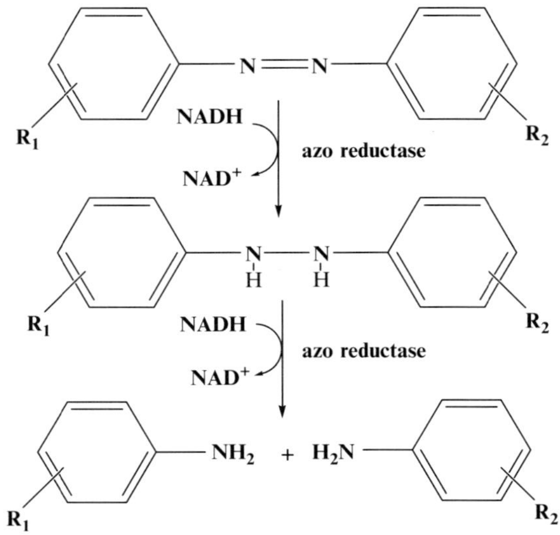

Fig. 2 The decolorization mechanism of azo dyes

2.2 Mediated Biological Azo Dye Reduction

As the aforementioned direct mechanism of azo dye biotical reduction, azo dye can be catalytically reduced by specialized enzymes called azo reductases or by non-specific enzymes. Azo reductases are present in bacteria that are able to grow using only azo dye as a carbon and energy source. However, now there is no clear evidence of anaerobic azo reductase. And nonspecific enzymes catalyze the reduction of a wide range of electron-withdrawing contaminants, including azo dyes [53]. Thus, an indirect/mediated reduction is probably the main mechanism of dye reduction (Fig. 1), in which the redox mediators, such as NADH, NAD(P)H, $FMNH_2$, $FADH_2$, and quinines, act as redox equivalents or coenzymes to accelerate the cleavage of the azo bond. Now there are more literatures on the role of redox mediators in azo bond reduction by bacteria under anaerobic conditions (Table 1).

As shown in Table 1, the addition of anthraquinone-2,6-disulphonate could also greatly enhance the decolorization of many azo dyes [21]. Riboflavin in catalytic amounts significantly accelerates the reduction of mordant yellow 10 by anaerobic granular sludge [26]. 1-Amino 2-napthol, one of the constituent amines of the azo dye, AO7, increased its decolorization rate, possibly by mediating the transfer of reducing equivalents [42]. It was reported that cell suspensions of *Sphingomonas* sp. strain BN6 grown aerobically in the presence of 2-naphthyl sulfonate (NS) exhibited a 10–20-fold increase in decolorization rate of an azo dye, amaranth, over those grown in its absence. Based on these observations, a mechanism was proposed for the mediated reduction of azo dyes by *S. xenophaga*. Recently, Chang et al. also showed that the addition of culture supernatants containing metabolites of a dye-decolorizing strain, *E. coli* strain NO_3, enhanced azo dye decolorization rates [43].

Table 1 The effect on the azo dyes decolorization by redox mediators with different microorganism

Mediator	Electron donor	Azo dye	Results[a]	References
AN AQDS	Sulfide	Acid Orange 7 0.25–0.3 mM	Max. 13× Max. 105×	[19]
FAD	NADPH NADPH	Acid Red 27	Without FAD: less than 5% reduction	[20]
Autoclaved AGS	Sulfide	Acid Orange 7 0.25–0.3 mM	Max. 10×	[19]
AQDS	VFA	Reactive Red 2	Max. 7×	[21]
AQDS	VFA	Acid Orange 7	+	[22, 23]
AQDS	VFA	Reactive Red 2	+	[21]
2-NS-metab	Glucose	Acid Red 27	6×	[24]
AQS, AQDS, LAW, LAP PLUM	Glucose	Acid Red 27	+	[25]
AQS, LAW, Henna leaves	Glucose	Acid Red 27	AQS, max. 10–15×; LAW, max. 7–12×; Henna leaves, max. 9× (z.o.)	[25]
RF	VFA	Mordant Yellow 10	Max. 2×	[26]
AQS, LAW	Glucose	Acid Red 27 Acid Orange 20 Acid Orange 7 Food Yellow 3 Acid Red 18 Food Red 17 Acid Red 14 Acid Red 1 Acid Yellow 23 Acid Black 1 Food Black 1	Average effect 26×; *S. xenophaga*: AQS about 4× more effective than LAW; *E. coli*: LAW about 37× more effective than AQDS	[25]
AQDS	VFA, Glucose/VFA, glucose, H_2	Hydrolyzed Reactive Red 2	VFA, 1.4×; Glucose/VFA, 1.6×, Glucose, 2.3×; H2, 2.4×	[27]
AQDS	Glucose/VFA	Hydrolyzed Reactive Red 2	+	[27]
PAC	VFA	Hydrolyzed Reactive Red 2	+	[28]
AQDS	VFA, Glucose/VFA	Hydrolyzed Reactive Red 2	VFA: 1.7×, 1.9×, 1.9×, 1.6×, and 1.7×, at 45, 55, 60, 65, and 75°C, resp.; Glucose/VFA: 2.0×, 2.6×, 2.4×, 2.1×, and 1.5×, at 45, 55, 60, 65, and 75°C, resp.	[27]
AQS	Glucose	Acid Orange 52 "Dye I" Acid Orange 7 "Dye III"	AO52, max. 3×; no additional effect RM at RM/ADN2.1; "Dye I," similar results as with AO52; AO7 and	[29]

(*continued*)

Table 1 (continued)

Mediator	Electron donor	Azo dye	Results[a]	References
			"Dye III," adverse effect of RM (not quantified)	
AQDS, AQS, RF, (CNB12)	Glucose/VFA	Reactive Red 2 Acid Orange 7 Mordant Yellow 10	+	[30]
AQDS	Glucose/VFA	Hydrolyzed Reactive Red 2	30°C, 5×; 55°C, 1.5×; k of RM-free control is 6× higher at 55°C as compared to 30°C	[31]
AQDS	Glucose/VFA	Hydrolyzed Reactive Red 2	1.4–1.7×	[31]
AQDS	Glucose/VFA	Hydrolyzed Reactive Red 2	Max. 5.9×	[31]
AQS	Sulfide	Reactive Red 2 0.3 mM	1.2×	[30]
AQS	Sulfide	Hydrolyzed Reactive Red 2 0.3 mM	1.3–1.5×	[30]
AQDS	VFA	Hydrolyzed Reactive Red 2, textile wastewater	+	[32, 33]
MQ, AQDS	SA	Acid Orange 7	+	[34]
AQS, (BQ), (AQS+BQ)	Glucose	Hydrolyzed Reactive Red 2	AQS: 3.8× and 2.3× at 30 and 55°C, resp.; BQ: slight stimulation at 30°C, slight adversary effect at 55°C; AQS +BQ: no effect; k of RM-free control is 5.6× higher at 55°C as compared to 30°C	[33]
RF	Glucose	Reactive Red 2 Reactive Orange 14	RR2, 1.7×; RR4, no stimulation; RO14, 2.9×	[32]
RF	Acetate, H_2, formate, methanol	Reactive Red 2	Acetate, 3.7× and 1.5×; H2, 4.6× and 3.8×; formate, 2.1× and 2.0×; methanol, 2.8× and 2.4×	[32]
RF	H_2	Reactive Red 2	5.3×; BES stimulates azo dye reduction	[32]
RF	Glucose	Reactive Red 2	Max. 23× and 6×	[32]
AQS, (BQ), (AQS+BQ)	Glucose	Hydrolyzed Reactive Red 2	AQS: 3.8× and 2.3×, at 30 and 55°C, resp.; BQ: slight stimulation at 30°C, slight adversary effect at 55°C; AQS +BQ: no effect; k of RM-free control is	[33]

(*continued*)

Table 1 (continued)

Mediator	Electron donor	Azo dye	Results[a]	References
			5.6× higher at 55°C as compared to 30°C	
AQDS, LAW, RF	Glucose	Reactive Orange 14 Direct Blue 53 Direct Blue 71	Stimulatory effect varies largely between different dye-RM combinations: avg. 2×; max. 3.8×	[35]
RF	Glucose	Reactive Orange 14	1.5–2×	[23]
AD-metab.	Glucose	Acid Orange 52	~2×	[35]
RF	Glucose	Reactive Red 2 Reactive Red 4 Reactive Orange 14	RR2, 1.7×; RR4, no stimulation; RO14, 2.9×	[3]
RF	Glucose at different conc.	Reactive Orange 14	1.4–4.0×	[3]
RF	Acetate, H_2, formate, methanol	Reactive Red 2 Reactive Orange 14	Acetate, 1.5× and 4.6×; H2, 3.8× to 7.1× (RR2 and RO14, resp.); formate, 2.0×; methanol, 2.4×	[3]
RF	Acetate	Acid Orange 6	+	[36]
RF	Sulfide	Reactive Orange 14 0.15 mM	31–45×	[23])
AQDS	Formate	Reactive Black 5	3×	[37]
RF	H_2	Reactive Red 2	+	[34]
AQS, AQDS	Formate	Acid Red 27	AQS: 1.75×; AQDS: 1.33×	[38]
AQS, AQDS	Formate	Acid Red 73 Acid Yellow 36 Acid Orange 10 Acid Orange 7 Acid Red 26	+	[38]
GAC (dye-saturated), graphite, aluminium	Acetate	Acid Orange 7	+	[39]
Graphite	Acetate	Acid Orange 7	+	[39]
Alginate beads with AQ	YE/peptone	Reactive Red 24	+	[40]
Different AQ compounds[i]	YE/peptone	Reactive Red 24	+	[6]
BA	YE/peptone	Reactive Red 24 Reactive Red 2 Acid Red 1 Acid Red 14 Acid Red 72 Acid Black 1	Max. 1.9×, 1.7×, 1.5×, 1.3×, 2.3×, 1.8× (z.o., RR24, RR2, AR1,AR14, AR72, AB1,	[6]
AQDS	Formate	Orange 5	~4× (30°C)	[41]

[a]× expresses the fold increase of zero-order reduction rates as compared to the RM-free controls; + expresses the increase but not the fold

Reduced flavins ($FADH_2$, $FMNH_2$, and riboflavin) generated by flavin-dependent reductases have been hypothesized to reduce azo dyes in a nonspecific chemical reaction, and flavin reductases have been revealed to be indeed anaerobic azoreductases. Other reduced enzyme cofactors, for example, NADH, NADH, NADPH, and an NADPH-generating system, have also been reported to reduce azo dyes. Except for enzyme cofactors, different artificial redox mediating compounds, especially such as quinines, are important redox mediators of azo dye anaerobic reduction (Table 1).

2.3 *Azo Dye Decolorization by Biogenic Inorganic Compounds*

During the azo dye production and application processes, many reducing chemical compounds are often added and may be transformed to biogenic reductants such as dithionite and zerovalent iron, as well as inorganic compounds such as sulfide and ferrous ion, as end products of metabolic reactions under anaerobic conditions [12–15]. These chemical reductants result in the chemical reduction of azo dyes under anaerobic conditions (Fig. 1). For example, sulfate is often an additive of dyebaths or it is formed by the oxidation of more reduced sulfur species used in dyeing processes, such as sulfide, hydrosulfite, and dithionite. Sulfate also results from the neutralization of alkaline dye effluents with sulfuric acid. Sulfide is therefore a relevant compound, as it will be generated by sulfate-reducing bacteria during treatment of these wastewaters in anaerobic bioreactors. H_2S generation by SRB has been proved to reduce azo dyes in the extracellular [44, 45]. There is a competition between sulfate and dye to become the terminal electron acceptor of the reducing equivalents. Van der Zee et al. observed that different sulfate concentrations did not have an adverse effect on the reduction of RR2 in either batch assays or reactor experiments [28].

3 Redox Mediator System

3.1 *Types of Redox Mediators*

In the field of the reductive (bio)transformation of priority pollutants, the reported redox mediator molecules include cytochromes, pyridines, cobalamins, porphyrins, phenazines, flavines, and quinines [12–15]. However, Quinones have been studied as the most appropriate RM for the reductive (bio)transformation of azo dyes [12].

Now there are many studies on the different redox mediators in azo bond reduction by bacteria under anaerobic conditions. The types of redox mediators are listed in Table 1.

3.2 Acceleration Mechanism of Redox Mediators

As discussed earlier, Azo biological decolorization are mainly reduced in a direct reduction or mediated/indirect reduction with nonspecial azo reductase or reduced enzyme cofactors (Figs. 1 and 3). According to the direct enzymatic reduction mechanism, nonspecial azo reductase can catalyze the transfer of reducing equivalents originating from the oxidation of original electron donor in the azo dyes. In

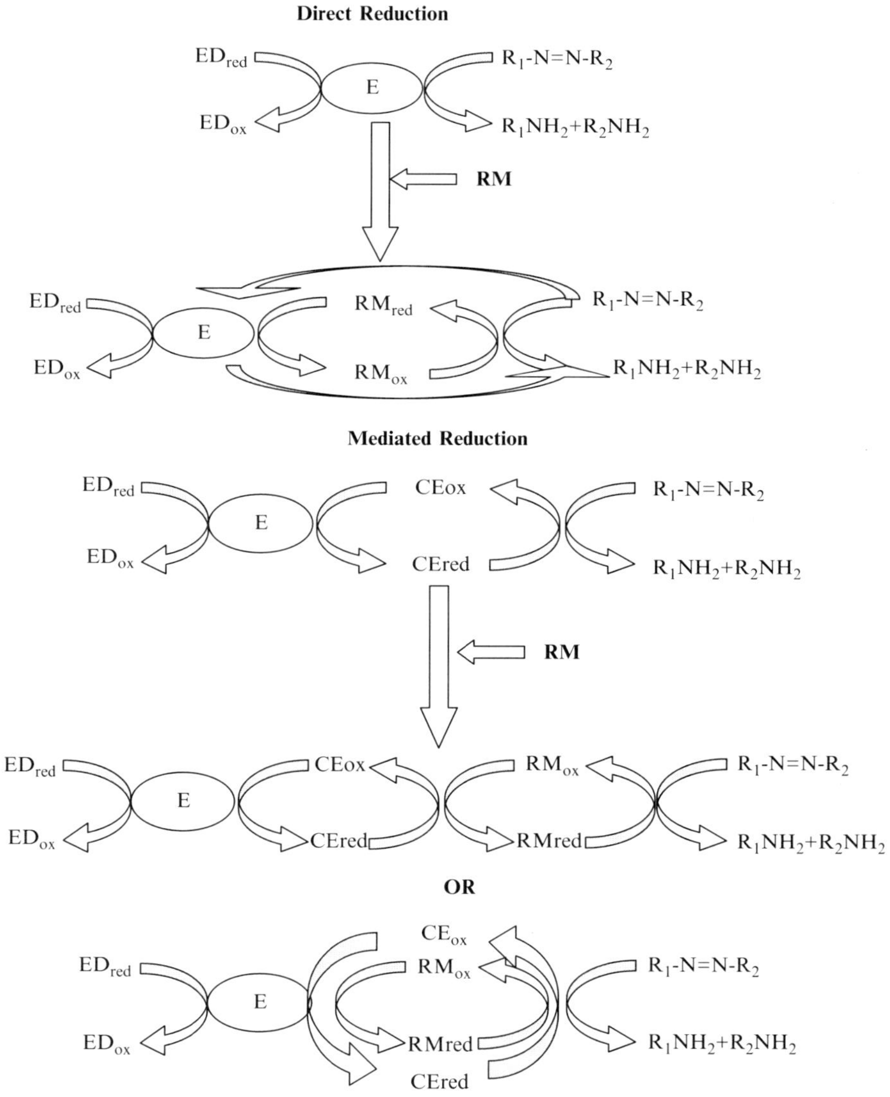

Fig. 3 The presumed accelerating mechanisms of redox mediators

light of the mediated/indirect reduction mechanism, azo dyes are reduced by azo reductase cooperated with coenzymes, such as $NADH/FADH_2$, which the oxidized and reduced state of coenzymes convert with the reduction process of azo dyes and the oxidation process of original electron donor.

The acceleration mechanism of redox mediators are presumed by van der Zee [15]. Redox mediators as reductase or coenzymes catalyze reactions by lowering the activation energy of the total reaction. Redox mediators, for example, artificial redox mediators such as AQDS, can accelerate both direct enzymatic reduction and mediated/indirect biological azo dye reduction (Fig. 3). In the case of direct enzymatic azo dye reduction, the accelerating effect of redox mediator will be due to redox mediator enzymatic reduction in addition to enzymatic reduction of the azo dye. Possibly, both reactions will be catalyzed by the same nonspecific periplasmic enzymes. In the case of azo dye reduction by reduced enzyme cofactors, the accelerating effect of redox mediator will either be due to an electron shuttle between the reduced enzyme cofactor and redox mediator or be due to redox mediator enzymatic reduction in addition to enzymatic reduction of the coenzymes. In the latter case, the addition of redox mediator simply increases the pool of electron carriers.

During the accelerating process, regeneration of redox mediator can be linked to the anaerobic oxidation of organic substrates by microorganisms.

However, the above presumed mechanism could not explain for all phenomena about redox mediator, which needs to be explored in details.

Theoretically, according to the mechanism of biological azo dye reduction, the processes of biological decolorization are oxidation–reduction reactions, in which transfer of electrons match with the proton flow by the help of coenzymes, such as NADPH/NADP+ and NADH/NAD+. The oxidation–reduction potentials of the couples of NADPH/NADP+ and NADH/NAD+ are –324 and –320 mV, respectively [25, 46]. The least $\Delta G_0'$ value of the conversion NADPH/NADP+ and NADH/NAD+ is 44 kJ [47]. Therefore, −93 mV, which is obtained from (1), could be considered as a rough limited ORP value for ordinary primary electron donors of the third mechanism of biological azo dye reduction. This was demonstrated by the results of many researches (Table 1). Hence, the observed failure of cyanocobalamin [30] and ethyl viologen [48] to act as a mediator is most probably due to their too low E_0' values: −530 and −480 mV, respectively.

$$\Delta G_0' = -2F\Delta E_0'(F = 96.6\text{kJ}/(\text{V mol})) \tag{1}$$

3.3 New Development of Redox Mediators Technology

It is known that the decolorization rate of azo dyes is increased by using redox mediators, which speed up the reaction rate by shuttling electrons from the biological oxidation of primary electron donors or from bulk electron donors to the electron-accepting azo dyes [21, 31, 40]. But continuous dosing of the dissolved redox mediators implies continuous expenses related to procurement of the chemical, as well as continuous discharge of this biologically recalcitrant compound. To take the accelerating effect of redox mediators on the azo dyes bio-transformation, several new developments of undissolved redox mediators technology are recently reported to overcome the limits of dissolved redox mediators technology. The aim of this section is to underline the new research fields to enhance the applicability of redox mediators in azo dyes decolorization.

Van der Zee et al. have reported that activated carbon, which is known to have quinone groups on its surface, enhanced dye decolorization [28]. This is probably one of the first examples of biocatalysis mediated by activated carbon. An AC-packed bioreactor enhances the decolorization rate higher than a bioreactor control lacking AC during the reductive decolorization of RR2. The results indicated that the redox mediating capacity of AC was the main cause of the enhanced decolorization. However, the accelerating effects of AC gradually decrease, which is attributed to its continuous wash-out from the reactor. Similar to the above study, Mezohegyi et al. achieved high decolorization rate of AO7 with an upflow packed-bed reactor (UPBR) containing biological AC [39].

Another undissolved redox mediators technology is reported to immobilize anthraquinone by entrapment in calcium alginate (CA), Polyvinyl alcohol (PVA)-H_3BO_3, and agar [40]. In this study, immobilized anthraquinone (AQ) with calcium alginate increase twofold the decolorization rate of different azo dyes by a salt-tolerant bacterial culture, compared to controls lacking AQ. The reusability of the anthraquinone immobilization beads was evaluated with repeated-batch decolorization experiments. After four repeated experiments, the decolorization rate of CA immobilized anthraquinone retained over 90% of their original value. The experiments explored a great improvement of the redox mediator application and the new bio-treatment concept. This immobilized redox mediator technology is also carried out by Su et al. [49]. However, the disadvantage of this technology is that the accelerating effect of redox mediator was lost gradually with the disruption of the polymeric material owing to weak mechanical strength of the materials explored.

Lately, the accelerating effect of functionalized polypyrrole (PPy) composites consisting of ACF/PPy/AQDS is studied during the biological decolorization processes of azo dyes [50]. This study suggests that ACF/PPy/AQDS play a good catalytic role and accelerate the reductive decolorization of different azo dyes by an anaerobic consortium. But their use in the practical full-scale wastewater treatment still needs to be proved in the future studies.

In a word, these studies explore a great improvement of the redox mediator application and the new bio-treatment concept for biological treatment.

4 Conclusion

The effects of redox mediators are different as reported in the present literatures. On the one hand, the accelerating effects of dissolved or undissolved redox mediators have been studied in details in the bio-decolorization processes in the above review.

On the other hand, the inhibitory effects are also discussed in several reports [51, 52]. However, there are few literatures about the exact and well catalytic mechanisms of dissolved or undissolved redox mediators, which are the bottlenecks of the accelerating/inhibitory effects, the fast development, and the more application of dissolved or undissolved redox mediators. Therefore, the catalytic mechanisms of dissolved or undissolved redox mediators are the focus for the anaerobic bio-transformation of priority pollutants in the future. At the same time, the more effective undissolved redox mediators is also another noticed field during the new anaerobic bio-technology of wastewater treatment.

References

1. Delee W, O'Neill C, Hawkes FR et al (1998) Anaerobic treatment of textile effluents: a review. J Chem Technol Biotechnol 73(4):323–335
2. Pandey A, Singh P, Iyengar L (2007) Bacterial decolorization and degradation of azo dyes. Int Biodeterior Biodegradation 59:73–78
3. Dos Santos AB, Bisschops IAE, Cervantes FJ (2006) Closing process water cycles and product recovery in textile industry: perspective for biological treatment. In: Cervantes FJ, Van Haandel AC, Pavlostathis SG (eds) Advanced biological treatment processes for industrial wastewaters, vol 1. International Water Association, London, pp 298–320
4. Talarposhti AM, Donnelly T, Anderson GK (2001) Colour removal from a simulated dye wastewater using a two phase anaerobic packed bed reactor. Water Res 35:425–432
5. Dos Santos AB, Cervantes FJ, van Lier JB (2007) Review paper on current technologies for decolourisation of textile wastewaters: perspectives for anaerobic biotechnology. Bioresour Technol 98:2369–2385
6. Guo J, Zhou J, Wang D et al (2008) The new incorporation biotreatment technology of bromoamine acid and azo dyes wastewaters under high salt conditions. Biodegradation 19:15–19
7. Weisburger JH (2002) Comments on the history and importance of aromatic and heterocyclic amines in public health. Mutat Res 506–507:9–20
8. Guo J, Zhou J, Wang D et al (2005) Decolorization of dye wastewater with high salt concentration by the Acclimatized salt-tolerant cultures. J Environ Sci 17(6):984–988
9. Hao OJ, Kim H, Chiang PC (2000) Decolourisation of wastewater. Crit Rev Environ Sci Technol 30:449–505
10. Robinson IM, Mcmullan G, Nigam P (2001) Remediation of dyes in textile effluent: a critical review on current treatment technologies. Bioresour Technol 77:247–255
11. Field JA, Cervantes FJ (2005) Microbial redox reactions mediated by humus and structurally related quinones. In: Perminova IV, Hatfield K, Hertkorn N (eds) Use of humic substances to remediate polluted environments: from theory to practice, vol 52. Springer, Dordrecht, pp 343–352
12. Van der Zee FP, Cervantes FJ (2009) Impact and application of electron shuttles on the redox (bio)transformation of contaminants: a review. Biotechnol Adv 27:256–277
13. Cervantes FJ (2002) Quinones as electron acceptors and redox mediators for the anaerobic biotransformation of priority pollutants. Ph.D. thesis, Wageningen University, Wageningen
14. Guo J (2006) Biodegradation of hyper-salinity dye wastewaters and the accelerating effect of redox mediators. Ph.D. thesis, Dalian Technology University, Dalian
15. Van der Zee FP (2002) Anaerobic azo dye reduction. Ph.D. thesis, Wageningen University, Wageningen

16. Blumel S, Stolz A (2003) Cloning and characterization of the gene coding for the aerobic azoreductase from *Pigmentiphaga kullae* k24. Appl Microbiol Biotechnol 62:186–190
17. Madigan MT, Martinko JM, Parker J (2003) Brock biology of microorganisms, 10th edn. Prentice-Hall, Upper Saddle River
18. Van der Zee FP, Bouwman RHM, Strik DPBTB et al (2001) Application of redox mediators to accelerate the transformation of reactive azo dyes in anaerobic bioreactors. Biotechnol Bioeng 75:691–701
19. Van der Zee FP, Lettinga G, Field JA (2000) The role of (auto)catalysis in the mechanism of anaerobic azo reduction. Water Sci Technol 42:301–308
20. Russ R, Rau J, Stolz A (2000) The function of cytoplasmic flavin reductases in the reduction of azo dyes by bacteria. Appl Environ Microbiol 66:1429–1434
21. Van der Zee FP, Lettinga G, Field JA (2001) Azo dye decolorization by anaerobic granular sludge. Chemosphere 44:1169–1176
22. Cervantes FJ, Dijksma W, Duong-Dac T et al (2001) Anaerobic mineralization of toluene by enriched sediments with quinones and humus as terminal electron acceptors. Appl Environ Microbiol 67:4471–4478
23. Cervantes FJ, Enriquez JE, Mendoza-Hernandez MR et al (2006) The role of sulphate reduction on the reductive decolorization of the azo dye reactive orange 14. Water Sci Technol 54:171–177
24. Keck A, Rau J, Reemtsma T et al (2002) Identification of quinoide redox mediators that are formed during the degradation of naphthalene-2-sulfonate by *Sphingomonas xenophaga* BN6. Appl Environ Microbiol 68:4341–4349
25. Rau J, Knackmuss HJ, Stolz A (2002) Effects of different quinoid redox mediators on the anaerobic reduction of azo dyes by bacteria. Environ Sci Technol 36:1497–1504
26. Field JA, Brady J (2003) Riboflavin as a redox mediator accelerating the reduction of the azo dye Mordant Yellow 10 by anaerobic granular sludge. Water Sci Technol 48:187–193
27. Dos Santos AB, Cervantes FJ, Yaya-Beas RE et al (2003) Effect of redox mediator, AQDS, on the decolourisation of a reactive azo dye containing triazine group in a thermophilic anaerobic EGSB reactor. Enzyme Microb Technol 33:942–951
28. Van der Zee FP, Bisschops IAE, Lettinga G et al (2003) Activated carbon as an electron acceptor and redox mediator during the anaerobic biotransformation of azo dyes. Environ Sci Technol 37:402–408
29. Ramalho PA, CardosoMH Cavaco-Paulo A et al (2004) Characterization of azo reduction activity in a novel ascomycete yeast strain. Appl Environ Microbiol 70:2279–2288
30. Dos Santos AB, Bisschops IAE, Cervantes FJ et al (2004) Effect of different redox mediators during thermophilic azo dye reduction by anaerobic granular sludge and comparative study between mesophilic (30°C) and thermophilic (55°C) treatments for decolourisation of textile wastewaters. Chemosphere 55:1149–1157
31. Dos Santos AB, Cervantes FJ, Van Lier JB (2004) Azo dye reduction by thermophilic anaerobic granular sludge, and the impact of the redoxmediator anthraquinone-2, 6-disulfonate (AQDS) on the reductive biochemical transformation. Appl Microbiol Biotechnol 64:62–69
32. Dos Santos AB, Bisschops IAE, Cervantes FJ et al (2005) The transformation and toxicity of anthraquinone dyes during thermophilic (55°C) and mesophilic (30°C) anaerobic treatments. J Biotechnol 115:345–353
33. Dos Santos AB, De Madrid MP, Stams AJM et al (2005) Azo dye reduction by mesophilic and thermophilic anaerobic consortia. Biotechnol Prog 21:1140–1145
34. Albuquerque MGE, Lopes AT, Serralheiro ML et al (2005) Biological sulphate reduction and redox mediator effects on azo dye decolourisation in anaerobic–aerobic sequencing batch reactors. Enzyme Microb Technol 36:790–799
35. Encinas-Yocupicio AA, Razo-Flores E, Sanchez-Diaz F et al (2006) Catalytic effects of different redox mediators on the reductive decolorization of azo dyes. Water Sci Technol 54:165–170
36. Yemashova N, Kalyuzhnyi S (2006) Microbial conversion of selected azo dyes and their breakdown products. Water Sci Technol 53:163–171

37. Pearce CI, Christie R, Boothman C et al (2006) Reactive azo dye reduction by *Shewanella* strain J18 143. Biotechnol Bioeng 95:692–703
38. Hong YG, Gu J, Xu ZC et al (2007) Humic substances act as electron acceptor and redox mediator for microbial dissimilatory azoreduction by *Shewanella* decolorationis S12. J Microbiol Biotechnol 17:428–437
39. Mezohegyi G, Kolodkin A, Castro UI, Bengoa C et al (2007) Effective anaerobic decolorization of azo dye Acid Orange 7 in continuous upflow packed-bed reactor using biological activated carbon system. Ind Eng Chem Res 46:6788–6792
40. Guo J, Zhou J, Wang D et al (2007) Biocalalyst effects of immobilized anthraquinone on the anaerobic reduction of azo dyes by the salt-tolerant bacteria. Water Res 41(2):426–432
41. Pearce CI, Guthrie JT, Lloyd JR (2008) Reduction of pigment dispersions by *Shewanella* strain J18 143. Dyes Pigm 76:696–705
42. Mendez-Paz D, Omil F, Lema JM (2005) Anaerobic treatment of azo dye acid Orange 7 under fed-batch and continuous conditions. Water Res 39:771–777
43. Chang JS, Chen BY, Lin YC (2004) Stimulation of bacterial decolorization of an azo dye by extracellular metabolites from *Escherichia coli* strain NO3. Bioresour Technol 91:243–248
44. Diniz PE, Lopes AT, Lino AR (2002) Anaerobic reduction of a sulfonated azo dye Congo Red by sulphate reducing bacteria. Appl Biochem Biotechnol 97:147–163
45. Yoo ES, Libra J, Wiesmannn U (2000) Reduction of azo dyes by *Desulfovibrio desulfuricans*. Water Sci Technol 41:15–22
46. Bourbonnais R, Leech D, Paice MG (1998) Electrochemical analysis of the interaction of laccase with lignin model compounds. Biochim Biophys Acta 1379:381–390
47. Zhu DX, Zheng CX (2003) Fundamentals of biochemistry. Science Press, Beijing, pp 523–527
48. Kudlich M, Keck A, Klein J (1997) Localization of the enzyme system involved in anaerobic reduction of azo yes by *Sphingomonas* sp. Strain BN6 and effect of artificial redox mediators on the rate of azo dye reduction. Appl Environ Microbiol 63:3691–3694
49. Su Y, Zhang Y, Wang J et al (2009) Enhanced bio-decolorization of azo dyes by co-immobilized quinone-reducing consortium and anthraquinone. Bioresour Technol. doi:10.1016/j.biortech.2009.01.029
50. Li L, Wang J, Zhou J et al (2008) Enhancement of nitroaromatic compounds anaerobic biotransformation using a novel immobilized redox mediator prepared by electropolymerization. Bioresour Technol 99:6908–6916
51. Okutman-Tas D, Pavlostathis SG (2007) The influence of iron reduction on the reductive biotransformation of pentachloronitrobenzene. Eur J Soil Biol 43:264–275
52. Tratnyek PG, Scherer MM, Deng BL (2001) Effects of natural organic matter, anthropogenic surfactants, and model quinones on the reduction of contaminants by zero-valent iron. Water Res 35:4435–4443
53. Stolz A (2001) Basic and applied aspects in the microbial degradation of azo dyes. Appl Microbiol Biotechnol 56:69–80

Bioreactors for Azo-Dye Conversion

Giuseppe Olivieri, Alberto Di Donato, Antonio Marzocchella, and Piero Salatino

Abstract This chapter embodies two sections. In the first section a survey of the state of the art of azo-dye conversion by means of bacteria is presented, with a focus on reactor design and operational issues. The relevance of thorough characterization of reaction kinetics and yields is discussed. The second section is focused on recent results regarding the conversion of an azo-dye by means of bacterial biofilm in an internal loop airlift reactor. Experimental results are analyzed in the light of a comprehensive reactor model. Key issues, research needs and priorities regarding bioprocess development for azo-dye conversion are discussed.

Keywords Biofilm, Bioreactor, Kinetics, Modeling, Reactor strategy

Contents

G. Olivieri, A. Marzocchella (✉), and P. Salatino
Department of Chemical Engineering, Università degli Studi di Napoli Federico II – School of Biotechnological Sciences, P.le V. Tecchio n. 80, 80125, Napoli, Italy
e-mail: giolivie@unina.it, marzocch@unina.it, salatino@unina.it

A. Di Donato
Department of Structural and Functional Biology, Università degli Studi di Napoli Federico II – School of Sciences, Complesso Universitario di Monte S. Angelo, Via Cinthia – 80126, Napoli, Italy
e-mail: didonato@unina.it

H. Atacag Erkurt (ed.), *Biodegradation of Azo Dyes*,
Hdb Env Chem (2010) 9: 101–131, DOI 10.1007/698_2009_52,

Abbreviations

A	Aerobic
AC	Actived carbon
AN	Anaerobic
B	Biofilm
BAC	Biological actived carbon
COD	Chemical oxygen demand
E	Entrapped cells
F	Free cells
GAC	Granular actived carbon
MLSS	Mixed liquor suspended solids
SBR	Sequential batch reactor
UASB	Upflow anaerobic sludge bed
VSS	Volatile suspended solids
WW	Waste-water

1 Survey of the State of the Art of Azo-Dye Bioconversion

The present survey addresses studies on azo-dye conversion by means of bacteria, with a close focus on bioreactor design and operational strategies. Other chapters of the Handbook and recently published reviews [1, 2] address instead the detailed biochemical pathways underlying azo-dye conversion and the fate of the conversion products.

Table 1 is a survey of studies in which azo-dye conversion is investigated, with an emphasis on process characterization and development. Process kinetics and reactor typologies, design and operation are reported. The ranges of pH, dyes' concentration and temperature investigated are also indicated in the table.

Dyes investigated in the studies listed in Table 1 may be grouped into three broad categories: acids (11 dyes, 26 papers), directs (5 dyes, 5 papers) and reactives (7 dyes, 13 papers). Acid orange 7 is by far the most investigated (12 papers). Though the spectrum of investigated dyes is quite wide, it is still a tiny fraction of the broad range of azo-dyes commonly used in industrial processes. Pearce et al. [1] highlighted that the dye's structure plays a key role in bioconversion. In particular, the color removal rate decreases with increasing molecular weight and structural complexity. This feature calls for more comprehensive understanding of the structure–reactivity relationships and quantitative assessment of conversion kinetics and yields for a broader selection of dyes.

Processes reported in Table 1 are typically anaerobic (AN). In agreement with the observations reported by Wuhrmann et al. [49], azo-dye bioconversion occurs with the standard organism and other facultative or obligatory aerobic bacteria in exclusively anoxic conditions. Different methods can be used to establish the required anaerobic conditions. A common procedure is simply sparging oxygen-free gas

Table 1 Contributes available in the literature on azo-dye conversion by means of bacteria reactors

References	Microrganism(s)	Dyes	Reactor	Modality	Reaction environment	Biophase[a]	Entrapment/ supports	Dye_L (mg/L)	T (°C)	pH
[3]	Mixed liquor from a municipal WW treatment plant	Acid orange 8 Acid orange 10 Acid red 14	Rotating drum biofilm reactor	Continuous	Aerobic	B	Drum	–	22	7
[4]	Consortium	Reactive red 141	Bottles	–	Anaerobic	–	–	100, 150, 200	–	–
[5]	Mixed liquor from a municipal WW treatment plant	Acid orange 8 Acid orange 7 Acid orange 10 Acid orange 14	Rotating drum biofilm reactor	Continuous	Aerobic	B	Drum	[COD = 60 – 750 mg/L]	22	7
[6]	Methanogenic consortium	Mordant orange 1	UASB	Continuous	Anaerobic	B	Sludge granules	50–200	30	n.a.
[7]	*Pseudomonas luteola*	Reactive red 22	Aerated flask	Batch and fed-batch	Aerobic growth, anaerobic decolorization	F	–	200–4,000	28	n.a.
[8]	Activated sludge from municipal WW treatment plant	Disperse blue 9	Biofilter	Batch	Sequential anaerobic/ aerobic	B	–	<120	–	–
[9]	Granules from a paper pulp processing plant	Red h-e7b	UASB and aerobic reactor	Continuous	Sequential anoxic and aerobic	B	Sludge granules	150–750	25	7.32
[10]	Mixed culture, methanogenic culture	Acid orange 7	Bottle	Batch	Anaerobic	B	Flocculent sludge	60–300	37	n.a.
[11]	*Pseudomonas luteola*	Reactive red 22	Flask	Sequential batch	Static and agitated incubation	E	Ca arginate, k carageenan polyacryl. gel	0–200	20–47	5–10
[12]	*Pseudomonas luteola*	Reactive red 22	Flask	Batch	Static incubation	F	–	0–400	20–47	5–9
[13]	Aerobic granular sludge	20 dyes	UASB	Batch	Anaerobic by N_2/ CO_2 spreading	B	GAC	0.3 mM	30	n.a.

(*continued*)

Table 1 (continued)

References	Microrganism(s)	Dyes	Reactor	Modality	Reaction environment	Biophase[a]	Entrapment/ supports	Dye_L (mg/L)	T (°C)	pH
[14]	*Pseudomonas st.s*	Acid violet 7 Acid red 151 Reactive black 5 Acid yellow 34	Flask	Batch	Static anoxic	F	–	100	35–40	5–10
[15]	*Sphigomonas* sp. *1CX* and *SAD4i*	Acid orange 7	Rotating drum	Continuous	Aerobic	B	Drum surface	25–290	28	n.a.
[16]	Sulfite by sulfate reducing bacteria	Reactive orange 96	Flask	Batch	Anaerobic	F	–	30–120	37	4.1–7.1
[17]	Consortium	Mixture	Rotating biological contactor	Continuous	Aerobic	B	Laterite stone	25–100	28	8.5–10.5
[18]	Consortium	Red rbn	Mechanical mixing	Batch and continuous	Anaerobic by nitrogen spreading	E	PVA gel beads (d = 3–4 mm)	50–500	30	4–10
[19]	*Sphigomonas* sp. *1CX* and *SAD4i*	Acid orange 7	Rotating drum	Continuous	Aerobic	B	On the drum surface	50–100	n.a.	7
[20]	*Escherichia coli Pseudomonas* sp.	Congo red direct black 38	Bottle	Batch	Aerobic, anaerobic, microaerophilic	F	–	250–3,000	30, 35	6.3–9.47
[21]	Two isolated strains and *Pseudomonas* sp.	Mix of 7 dyes	Single-stage packed bed	Continuous	Aerobic	B	Laterite stone	25–100	30	7–9
[22]	Activated sludge from a municipal WW treatment plant	Acid red 151	Fixed bed	Sequencing batch	Aerobic	B	Porous volcanic rock (puzolane)	25, 150	25	7
[23]	Consortium	Reactive black 5 direct brown 2	Shaked bottles	Batch	Anaerobic	B	Granulated anaerobic sludge	200–3,200	35	6.6–7.05

[24]	Activated sludge from a municipal WW treatment plant	Acid orange 7	Rotating drum biofilm	Batch and continuous	Aerobic	B	Drum	15–50	25	7.7–8
[25]	*Enterobacter agglomerans*	Methyl red	Flask	Repeated batch	Aerobic	E	Ca arginate, polyacrylamide gel, vermiculite, Cu beech	100	25	7
[26]	*Pseudomonas luteola*, modified *E.coli*	Reactive black b	Flask	Batch	Aerobic growth, anaerobic conv.	F	–	0–1,500	15–50	4–10
[27]	Mixed, mesophilic methanogenic culture	Reactive red 1	Bottles	Batch	Methanogenic conditions	F	–	50–2,000	35	7
[28]	Anaerobic sludge from a full-scale UASB	Acid orange 7 direct red 254	UASB	Continuous	Methanogenic conditions	B	Sludge granules	60–1,800	37	n.a.
[29]	*Pseudomonas luteola*	Reactive red 22	Fixed bed	Continuous	Anaerobic	E	Ca alginate, polyacrylamide gel	30–200	n.a	n.a.
[30]	Biomass from an anaerobic digester	Reactive dyes disperse dyes	Fixed bed with recycle and fixed bed	Continuous	Anaerobic	B	Reticulated sintered glass	n.a.	37	6.4–7.2
[31]	Anaerobic sludge from UASB	Reactive red 2	Bottle	Discrete continuous reactor	Anaerobic	B	Sludge	100, 200	27–29	7.2–7.5
[32]	Sludge from a municipal WW treatment plant	Orange II	SBR system	24 h sequencing batch	Aerobic/anaerobic	B	GAC	100	25	n.a.
[33]	Sludge from a municipal WW treatment plant	Orange II	UASB and SBR	24 h sequencing batch	Anaerobic and aerobic	B	Sludge	60–300	16–30	n.a.

(*continued*)

Table 1 (continued)

References	Microrganism(s)	Dyes	Reactor	Modality	Reaction environment	Biophase[a]	Entrapment/ supports	Dye_L (mg/L)	T (°C)	pH
[34]	Sludge from a municipal WW treatment plant	Orange II	Aerobic and anaerobic SBRs	Sequencing batch	Aerobic/anaerobic	B	Sludge	50, 100	25	n.a.
[35]	Bacterial consortium	Dye mixture	Microaerophilic reactor + aerobic reactor	Continuous	Anaerobic/aerobic	B	Beads used for insulation	56	Room	7.2
[36]	*Aeromonas hydrophila, Comamonas testosterone, Acinetobacter baumannii*	Red rbn	Fluidized bed	Continuous	Anaerobic	E	PVA gel beads	0–4,400	30	n.c.
[37]	Methanogens consortium	Remazol black-b, rem. red rr, rem. yellow rr	Fluidized bed	Continuous	Anaerobic	B	Sinter glass beads	60	37	6.4–7.2
[38]	Sludge from methanogenic reactor	Acid blue 113, direct black 22, sarasit blue sr	UASB and aerobic CSTR	Continuous	Anaerobic, aerobic	B	Granulated anaerobic sludge	200	n.a.	6.9–7.3
[39]	Consortium	Acid red 88	Upflow fixed-film column and CSTR	Continuous	Anoxic and aerobic bioreactors	B	Polyurethane foam	20–100	20–45	6–7
[40]	Activated sludge from a full-scale plant	Remazol brilliant violet Remazol black b	SBR	24 h sequencing batch	Anaerobic/aerobic	B	Sludge	<100	30	n.a.
[41]	*Pseudomonas* sp. OX1	Acid orange 7	Flask, airlift	Batch	Aerobic growth, anaerobic conv.	F	–	10–1,000	25	6.9
[42]	Partial anaerobic digestion of aerobic sludge	Acid orange 7	Upflow packed bed reactor	Batch and continuous	Anaerobic by helium spreading	B	Alumina, graphite flakes, AC	50–100	35	6.8

[43]	Anaerobic sludge from a full scale UASB plant	Acid orange 6 acid orange 7	Fixed bed	Batch with biophase recycle	Sequential fixed-film anaerobic batch reactor	B	Ceramic rings	50–300	30–35	7.5
[44]	Consortium	Remazol black-b	Anoxic–oxic reactors	Continuous	n.a.	F	–	25–2,000	20–50	5–10
[45]	*Pseudomonas luteola*	Reactive red 22	Fixed bed	Continuous	Anaerobic	B	BAC	100–270	28	7
[46]	Consortium	Acid orange 7	Upflow packed bed reactor	Continuous	Anaerobic	B	BAC	100–300	35	6.7–7.4
[47]	Consortium	Acid orange 7	Packed column	Continuous	Anaerobic	B	GAC	<1,150	25	–
[48]	Consortium	Acid orange 7	Packed bed with liquid recirculation	24 h sequencing batch	Anaerobic	B	GAC	125–625	25	n.a.

[a] *F* Free cells; *E* entrapped cells; *B* biofilm

(helium, nitrogen, carbon dioxide, etc.) into the reactor. Anaerobic conditions of static cultures may also be established as oxygen depletion is promoted by the respiration of facultative bacteria. Addition of reducing agents (e.g., sodium thioglycollate) to the culture has been occasionally used to promote the establishment of anaerobic conditions.

Some attempts to convert azo-dyes under aerobic conditions in sludge/biofilm reactors have also been reported [15, 17, 19, 20, 34]. It is often hypothesized that azo-dye conversion is still carried out under microaerophilic conditions that establish in the inner region of the biophase, in agreement with the findings of Zhang et al. [5]. Typically, the aerobic stage results in a successful degradation of the products from the anaerobic azo-dye conversion. The reader is addressed to the review by Van der Zee and Villaverde [2] to learn more about the proper combination of anaerobic–aerobic stages in azo-dye treatment. Uncertainties on the fate of aromatic amines produced during the anaerobic conversion, a key issue for the success of the process, are addressed and discussed in their study.

Coupling between aerobic and anaerobic stages is also established through the production – which is typically associated with bacterial catabolism – of the reducing equivalents necessary to convert azo-dyes. It has been shown that the nature of the carbon source has a limited effect on azo-dye conversion, provided that the production rate of the reducing equivalents is large enough to sustain azo-dye conversion [6, 28]. Under batch conditions, the dye conversion degree starts to decrease when the initial carbon source decreases below a threshold value [7, 10, 20, 41, 43]. The aerobic–anaerobic stages coupling is better illustrated in Fig. 1, taken from Lodato et al. [41]. In this figure, the maximum extent of decolorization of acid orange 7 by *Pseudomonas* sp. OX1 in the anaerobic phase, or "decolorization quotient," is successfully correlated with the amount of carbon substrate converted during the preceding aerobic phase, regardless of the carbon source.

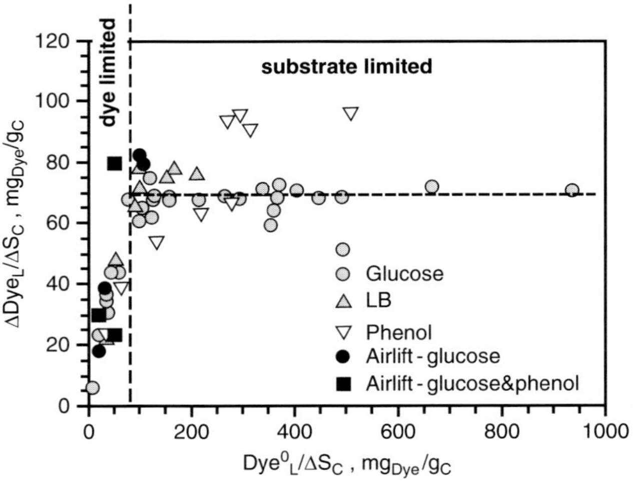

Fig. 1 Map of dye conversion limiting regimes: the "decolorization quotient" [41]

These findings suggest that a prerequisite for the successful continuous operation of bioreactors for azo-dye conversion is the thorough control of carbon loading: the azo-dye conversion degree drops to vanishing levels if carbon depletion occurs.

Bacteria responsible for azo-dye conversion can be found in different aggregation states: free cells, activated sludge, entrapped cells, biofilm on granular carriers or rotating surfaces. The biophase structure strongly affects the conversion process as regards both proper reactor selection and effective conversion kinetics. The first issue is discussed in the section "Bioreactor design and operational strategy for azo-dye conversion," and the second in the section "Reactor modeling for azo-dye conversion."

Studies carried out with free or entrapped cells have greatly contributed to the characterization of basic phenomena involved in dye conversion. However, from the practical standpoint, scale-up of processes based on either free or entrapped cells is not economically feasible. In fact, the first choice is usually associated with prohibitively large reaction volumes. The second choice is typically expensive and asks for industrial wastewaters of strongly controlled composition for the stability of entrapment matrixes to be preserved.

Bacteria used in azo-dye conversion are typically consortia. Only a few studies address single strains. Consortia were harvested in wastewater treatment plants, municipal or industrial. Some consortia were harvested from aerobic reactors, though most of the experience relates to anaerobic consortia. The single strains (*Pseudomonas*, *Sphigomonas*, *Escherichia coli*, *Bacillus cereus*, etc.) were typically isolated in bioremediation plants operating with xenobiotic-bearing wastewaters.

Some studies have demonstrated the ability of mixed bacterial cultures to promote azo-dye conversion. The main advantage of mixed cultures is the ability to cope with conversion of both the dye and the main products of dye degradation. Characterization of individual strains in consortia has seldom been accomplished during continuous tests. Only a few attempts have been made to characterize the morphology of bacteria during continuous operation of bioreactors and to relate process efficiency to the bacterial population [32, 33].

Processes reported in Table 1 have been carried out at temperature ranging, typically, between 20 and 40°C. Though the investigations carried out at lower temperature are very few [26, 33], this issue holds a key role in the design and optimization of the conversion processes. Provided that the heating-up of the wastewater streams is not economically feasible, the remediation process should be carried out at low temperature, particularly pressing in rigorous climate countries [50].

2 Bioreactor Design and Operational Strategy for Azo-Dye Conversion

Studies in Table 1 refer to different types of reactors: stirred tank (mechanically agitated reactors and rotary reactors), fixed bed and fluidized bed. Figure 2 reports an outline of the most common reactor typologies. The reactors may be operated

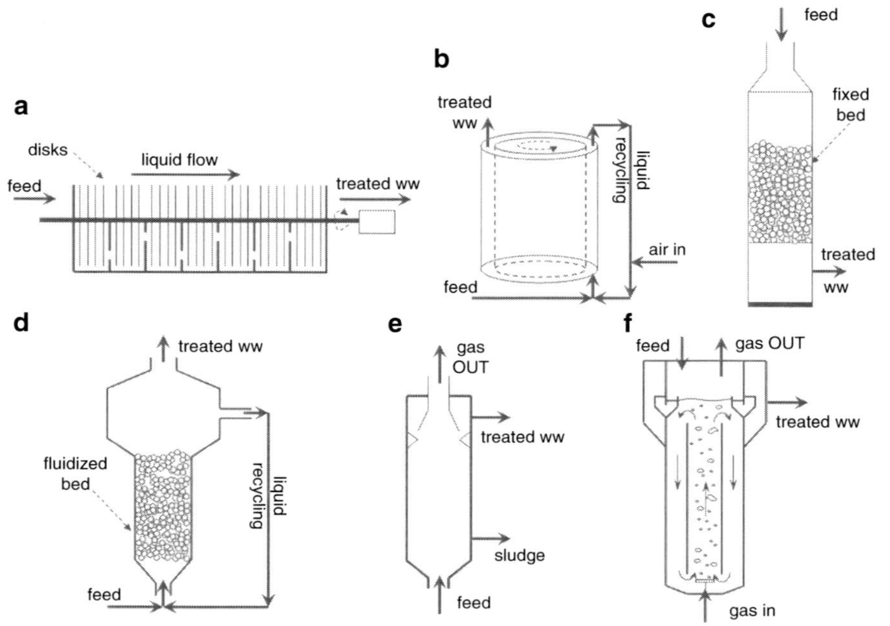

Fig. 2 Sketch of some reactor typologies used in azo-dye conversion. (**a**) rotating biological contactor; (**b**) drum reactor; (**c**) fixed bed reactor; (**d**) fluidized bed; (**e**) UASB; (**f**) airlift

under batch, sequential-batch and continuous conditions. The behavior of continuously operated reactors may span from uniform mixing to plug flow pattern. Typically, the behavior of a given reactor may change between the recalled extremes by tuning some of the operating conditions (e.g., recycle ratio).

A brief theoretical framework of the reactors will hereby be proposed to support the analysis of these reactors. The reader is addressed to textbooks [51, 52, 68] for a detailed treatment of the subject.

Stirred tank reactor (*STR*). The differential mass balance referred to the azo-dye converted by bacteria (assuming unstructured model for the biophase, i.e., that it is characterized only by cell mass or concentration *X*) yields

$$t = \int_{\mathrm{Dye_L}}^{\mathrm{Dye_L^0}} \frac{1}{X} \frac{\mathrm{d\,Dye_L}}{r_{\mathrm{dye}}(\mathrm{Dye_L}, c_{\mathrm{products}}, \mathrm{pH}, \ldots)}, \tag{1}$$

where t is the reaction time, $\mathrm{Dye_L^0}$ is the initial dye concentration, $\mathrm{Dye_L}$ is the dye concentration at time t in the reactor and r_{dye} is the dye conversion rate for unit of biophase mass (see next section for details) at the local conditions in the reactor. The reaction time depends strongly on the reaction rate and in particular on how metabolites accumulate in the reaction volume. The productivity is

$$W_{\text{dye}} = \frac{V\left(\text{Dye}_{\text{L}}^{0} - \text{Dye}_{\text{L}}\right)}{t + t_{\text{d}}}, \tag{2}$$

where V is the reactor volume and t_{d} is the dead time between two successive batches.

Continuous Stirred Tank Reactor (CSTR). The conversion degree of the azo-dye, the reaction volume (V) and the volumetric flow rate (Q) of the dye-bearing stream are related to each other through the material balance referred to the dye and extended to the reactor volume. Assuming an unstructured model for the biophase, the material balance yields

$$V = \frac{Q}{X} \frac{\left(\text{Dye}_{\text{L}}^{0} - \text{Dye}_{\text{L}}\right)}{r_{\text{dye}}\big|_{\text{out}}}, \tag{3}$$

where Dye_{L} is the dye concentration in the reactor and $r_{\text{dye}}\big|_{\text{out}}$ is the dye conversion rate for unit of biophase mass (see next section for details) at the conditions established in the reactor (concentration of the dye and of the products, pH, etc.).

Plug flow reactor (PFR) with recycle. The recycle reactor is characterized by a non-zero value of R, that is the ratio between the mass flow rate of the recycled stream and the feeding rate Q. The material balance reads for this case as

$$V = (R+1)Q \int_{\text{Dye}_{\text{L}}}^{\frac{\text{Dye}_{\text{L}}^{0} + R \times \text{Dye}_{\text{L}}}{R+1}} \frac{1}{X} \frac{d\text{Dye}_{\text{L}}}{r_{\text{dye}}\left(\text{Dye}_{\text{L}}, c_{\text{products}}, \text{pH}, \ldots\right)}, \tag{4}$$

where the integral extends over the span of dye concentrations. In (4) the axial profile of the reaction environment is reflected by the axial profile of r_{dye}.

The PFR without recycle is described in (4) just by setting $R = 0$. Equation (4) yields (2) for R tending to infinity.

The productivity of continuously operated reactors at steady state is

$$W_{\text{dye}} = Q\left(\text{Dye}_{\text{L}}^{0} - \text{Dye}_{\text{L}}\right), \tag{5}$$

where Dye_{L} is the dye concentration in the treated stream.

Equations (1), (3) and (4) should be coupled with a balance on the biophase to determine the value of X. Under conditions characterized by hindered growth, the biophase concentration may be assumed constant in both equations. In practice, this implies effective biophase containment in the reactor.

The sequential batch reactor (SBR) consists of a vessel operated under batch conditions according to the time schedule reported in Fig. 3. The symbols Fill, React, Settle, Draw and Idle refer to the typical sequential phases of operation: loading, reaction, biophase settling, discharging and the idle time. The reaction period may be split into two sub-phases: an anaerobic phase and an aerobic phase. The aerobic sub-phase is devoted to convert products of the azo-dye anaerobic

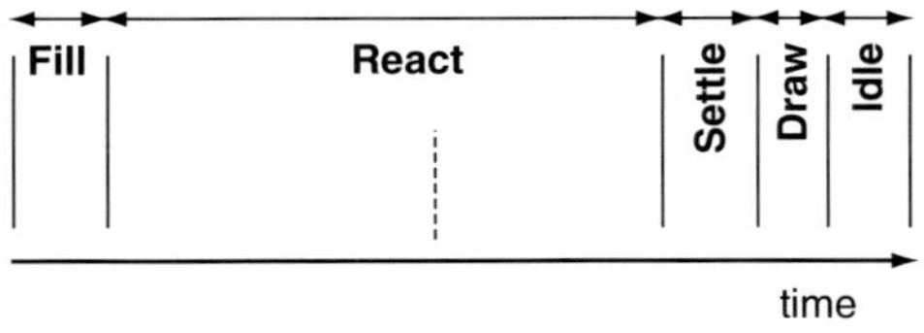

Fig. 3 Time schedule program of a typical sequencing batch reactor

conversion [32–34, 40]. The dye conversion measured in the discharged solution depends on the time evolution of the conversion process with the progressive reduction of the dye and the accumulation of the products, even during the Settle and Draw phases. Behavior of the reactor is described by means of the STR models. The productivity may be estimated by means of (2), assuming the cycle time at the denominator.

Performances and differences of the reactors studied and reported in Table 1 may be interpreted by means of the appropriate material balance, (1)–(4), and by considering the increase in the conversion rate with $\mathrm{Dye_L}$.

Hydraulic residence time (HRT $= V/Q$*).* Whatever be the ideal configurations of the continuous reactor, mass balances suggest that, for a given reactor and amount of biophase (V and X fixed), the conversion of the azo-dye should increase when Q decreases. In other words, $\mathrm{Dye_L}$ at the reactor exit should decrease when the HRT increases. Along with this effect, the observation of Senan et al. [21], Ong et al. [33], Bras et al. [28], Georgiou and Aivasidis [37] and the relations reported by van der Zee and Villaverde [2] may be interpreted.

Influence of the azo-dye concentration in the feeding depends on the reactor type and on the functional form of the reaction kinetics.

STR. The increase in $\mathrm{Dye_L^0}$ may be associated to a faster initial conversion if linear dependence of r_{dye} on $\mathrm{Dye_L}$ is active. However, the time to reach a pre-set final concentration may be even longer than that experienced at a lower $\mathrm{Dye_L^0}$ if products inhibition and/or substrate inhibition are active.

CSTR. Once the final dye concentration is set, the increase in the $\mathrm{Dye_L^0}$ may affect the reaction volume directly – increase of the concentration difference between the inlet and outlet of the reactor – and indirectly through the increase of product concentration in the reactor effluent. The increase of $\mathrm{Dye_L^0}$does not affect directly the conversion rate.

PFR. The reaction volume increases indefinitely as $\mathrm{Dye_L^0}$ is increased, for a given value of the final dye concentration to an extent that depends on the possible product-inhibited character of the reaction kinetics.

3 Kinetics of Azo-Dye Bioconversion

The kinetics of the azo-dye conversion plays a fundamental role in the design and operation of bioreactors. Two methodologies can be followed to assess the parameters of reaction kinetics [51, 52]: conversion rate estimated at the beginning of batch tests and conversion rate estimated during steady state continuous tests.

The mass balance on dye in a STR operated batchwise assuming an unstructured model yields

$$r_{\mathrm{dye}} = -\frac{1}{X}\frac{\mathrm{dDye_L}}{\mathrm{d}t}. \tag{6}$$

In agreement with (6), regression of dye concentration data measured at the beginning of the test makes it possible to relate the dye conversion rate to the conditions set in the reactor at the beginning of the test. Changing the initial conditions of the tests enables the evaluation of kinetic parameters.

With reference to a CSTR, working out (3) results in

$$r_{\mathrm{dye}} = \frac{D}{X}\left(\mathrm{Dye_L^0} - \mathrm{Dye_L}\right), \tag{7}$$

where $D(=Q/V)$ is the dilution rate. Equation (7) reports the dye conversion rate at the conditions established in the reactor (dye and products) under steady state conditions. Again, changing the operating conditions of the reactor makes it possible to determine the kinetic parameters.

Conversion rate data obtained under a wide range of operating conditions may be worked out to provide a kinetic expression, most typically expressed according to well established models for bioprocess kinetics: first and second order, Monod, Haldane, product-inhibited, etc.

The assessment of reaction kinetics by means of batch tests may be strongly affected by dye adsorption on the biophase and supports. The relevance of the adsorption phenomena of dyes on biophase has been addressed in studies regarding free cells [41], granular support biofilm [24], entrapped cells [11, 18], anaerobic sludge [10, 24, 31, 34] and biological activated carbon (BAC) [42, 45, 47, 48]. They have pointed out that the kinetics may be overestimated if the assessment of the adsorption contribution to the dye removal is not taken into account. Under batch conditions, the dye is fastly split between the liquid phase and the biophase, resulting in a sharp reduction of the dye concentration in the liquid phase until adsorption equilibrium is approached. The rate of dye adsorption must be estimated and ruled out in the kinetic assessment.

Experiments aimed at the characterization of the conversion kinetics under continuous reactor operation are not affected by adsorption phenomena. At steady state, the uptake of dye due to adsorption is practically zero since the biophase and supports are both in equilibrium with the liquid phase [53].

The relevance of species diffusion into the entrapped cells to the apparent conversion kinetics has been addressed by Chen et al. [54]. The authors have carried out a systematic study of the effects of beads' diameter and entrapped cell concentration on conversion rate.

Table 2 reports kinetic data and expressions available in the literature regarding azo-dye conversion. Unfortunately, data may not be compared directly with

Table 2 Kinetics data of azo-dye conversion

References	Dye	Biophase	Conversion rate [$mg_{dye}/(min\ g_{DM})$] with Dye_L in mg/L
[4]	Reactive red 141	Consortium, F	$r_{dye} = 7.3 \times 10^{-3}\ Dye_L\ mg_{dye}/L$ min, at $Dye_L^0 = 100$ mg/L $r_{dye} = 5.3 \times 10^{-3}\ Dye_L\ mg_{dye}/L$ min, at $Dye_L^0 = 150$ mg/L $r_{dye} = 4.2 \times 10^{-3}\ Dye_L\ mg_{dye}/L$ min, at $Dye_L^0 = 200$ mg/L
[9]	PROCION red h-e7b	Consortium, B	$r_{dye} = 4.0 \times 10^{-2}\ mg_{dye}/L$ min, at $Dye_L = 150$ mg/L
[11]	Reactive red 22	*P. luteola* – E	
		Polyacrylamide	$r_{dye} = 0.84\ Dye_L/(1{,}000 + Dye_L)$
		Ca alginate	$r_{dye} = 0.15\ Dye_L/(107 + Dye_L)$
		k-carrageenan	$r_{dye} = 0.18\ Dye_L/(267 + Dye_L)$
[12]	Reactive red 22	*P. luteola* – F	$r_{dye} = 0.20\ Dye_L/(156 + Dye_L)$
[13]	22 azo-dyes	Consortium	$r_{dye} = k\ Dye_L$, k ranges between 1 and 100 h, at $Dye_L = 0.3$ mM
[14]	Acid violet 7	*Pseudomonas* GM3, F	$r_{dye} = 0.12\,(Dye_L)^{0.5}$
[17]	Mixture of 7 azo-dye	Consortium – rotating biological contactor	0.48–186.71 $\mu g_{dye}/day\ (m_{disk\ surface})^2$
[18]	Red rbn	Consortium – E, polyvinyl alcohol	$r_{dye} = 14\ Dye_L/(196 + Dye_L)$
[23]	Reactive black 5	Consortium – B	$r_{dye} = 1.2 \times 10^{-3}\ Dye_L$
	Direct brown 2		$r_{dye} = 6.8 \times 10^{-4}\ Dye_L$
[25]	Methyl red	*Enterobacter agglomerans* – F	$r_{dye} = 8.2 \times 10^{-2}$ at $Dye_L = 100$ mg/L
		E, alginate	$r_{dye} = 5.1 \times 10^{-2}$, at $Dye_L = 100$ mg/L
		E, polyacrylamide	$r_{dye} = 2.1 \times 10^{-2}$, at $Dye_L = 100$ mg/L
		B, cooper beech	$r_{dye} = 1.6 \times 10^{-2}$, at $Dye_L = 100$ mg/L
		B, vermiculite	$r_{dye} = 1.5 \times 10^{-2}$, at $Dye_L = 100$ mg/L
[26]	Reactive black b	*Escherichia coli* NO3, F	$r_{dye} = 1.7\ Dye_L/(263 + Dye_L)$
		Escherichia coli CY1, F	$r_{dye} = 0.8\ Dye_L/(1{,}279 + Dye_L)$
[27]	Reactive red 2	Consortium, F	$r_{dye} = 1.8 \frac{Dye_L}{8.6 + Dye_L + Dye_L^2/7285}$
[29]	Reactive red 2	*P. luteola* – E, alginate	$r_{dye} = 6.9 \times 10^{-2}\ Dye_L/(69 + Dye_L)$
		Polyacrylamide	$r_{dye} = 4 \times 10^{-4}\ Dye_L$
[31]	Reactive red 2	Consortium, B	$r_{dye} = 1.2 \times 10^{-1}\ mg_{dye}/g_{VSS}$ min, at $Dye_L^0 = 100$ mg/L $r_{dye} = 6.5 \times 10^{-2}\ mg_{dye}/g_{VSS}$ min, at $Dye_L^0 = 200$ mg/L
[34]	Orange II	Consortium, B	$r_{dye} = 1.1 \times 10^{-6}\ mg_{dye}/g_{MLSS}$ min, (aerobic) and $r_{dye} = 5.2 \times 10^{-6}\ mg_{dye}/g_{MLSS}$ min, (anaerobic) at $Dye_L^0 = 50, 100$mg/L

(*continued*)

Table 2 (continued)

References	Dye	Biophase	Conversion rate [$mg_{dye}/(min\ g_{DM})$] with Dye_L in mg/L
[36]	Red rbn	*Aeromonas hydrophila, Comamonas testosterone, Acinetobacter baumannii* – F	$r_{dye} = 4.0\ Dye_L/(340 + Dye_L)$
		E, polyvinyl alcohol	$r_{dye} = 2.6\ Dye_L\ /(350 + Dye_L)$
[39]	Acid red 88	*Stenotrophomonas* sp., *Pseudomonas* sp. *and Bacillus* sp. – B, polyurethane foam	$r_{dye} = 0.13\ mg_{dye}/L$ min at $Dye_L = 2$–100 mg/L
[40]	Remazol brilliant violet	Consortium, B	$r_{dye} \cong 1.3 \times 10^{-2} Dye_L$, $Dye_L < 100$ mg/L[a]
	Remazol black b		$r_{dye} \cong k\, Dye_L$, k increases with Dye_L
[41]	Acid orange 7	*Pseudomonas* sp. OX1 – F	$r_{dye} = 0.11\ Dye_L/(290 + Dye_L)$
[43]	Acid orange 6	Consortium, B	$r_{dye} = 7.9 \times 10^{-2}\ mg_{dye}/L$ min at $Dye_L = 10$–200 mg/L[a] $r_{dye} = 1.2 \times 10^{-1}\ mg_{dye}/L$ min at $Dye_L = 18$–300 mg/L[a]
	Acid orange 7		$r_{dye} = 9.7 \times 10^{-3}\ c_{dye}\ (mg_{dye}/L\ min)$ at $Dye_L^0 = 3, 100$ mg/L
[45]	Reactive red 22	*P. luteola* – B, AC	$r_{dye} = 5.7\ Dye_L/(300 + Dye_L)$
[46]	Acid orange 7	Consortium – B, AC	$r_{dye} = 3.8 \times 10^4 \frac{Dye_L}{377 + Dye_L + Dye_L^2/14{,}370}$ mg_{dye}/L min

[a]Conversion rate of the investigated reactor: fixed bed operated as a plug flow reactor

each other since they refer to systems differing as regards the nature of the azo-dye and of the microrganism, extra carbon/energy source, kinetic assessment procedure.

Analysis of the equations/data reported in Table 2 highlights that the specific conversion rate depends strongly on both the selected azo-dye and the biophase. Assuming a concentration of 100 mg/L of a generic azo-dye, the specific conversion rate ranges between 10^{-2} and 5 $mg_{dye}/(g_{cell}$ min) with many data gathered around 0.1 $mg_{dye}/(g_{cell}$ min). It results that the order of magnitude of the specific conversion rate is quite small when estimated with reference to typical azo-dye concentrations in industrial wastewaters and at the very restrictive values set by environmental guidelines. In fact, for $Dye_L < 100$ mg/L, a quasi-linear relationship may be assumed for all kinetics reported in Table 2.

The thermal sensitivity of the specific conversion rate has been investigated by [12, 14, 26, 44]. Yu et al. [14], investigating a conversion process based on *Pseudomonas* sp. GM3, assumed that the k factor of the kinetic equation $r_{dye} = k(Dye_L)^{0.5}$ (see Table 2) increases with the temperature in agreement with the Arrhenius equation. They have also estimated the activation energy (16.87 kcal/mol) for the investigated process in the temperature range 10–35°C. Chang et al. [12] found a maximum of r_{dye} at 45°C and a fast deactivation of *P. luteola* cells at temperature larger than 45°C. Yeh and Chang [26] have investigated the temperature effects on r_{dye} by *E. coli* CY1 and *E. coli* NO3. They found that r_{dye} increases with T and a sharp decrease occurred at 37°C and 45°C for *E. coli* CY1 and *E. coli* NO3, respectively. Dafale et al. [44] investigated the effect of temperature on activated bacterial consortia and found a maximum r_{dye} at 37°C. Chang et al. [12] suggested that the loss of cell viability or the denaturation of the azo-reductase enzymes may be responsible for the observed decrease of r_{dye}.

An order-of-magnitude assessment of the biophase loading and bioreactor size needed to treat a reference wastewater stream is hereby presented. Based on the assumptions that the reactor is a continuous stirred bioreactor with a dye concentration at the inlet of $Dye_L^0 = 100$ mg/L, 90% conversion ($Dye_L = 10$ mg/L) and a specific conversion rate of order 10^{-2} $mg_{dye}/(g_{cell}$ min), the mass of the biophase needed to treat 0.1 m^3/min (about 150 m^3/day, a typical industrial wastewater stream) is about 1 kg. Assuming a biophase concentration of about 1 kg/m^3, a reactor volume of about 1 m^3 would be necessary to accomplish the proposed process. The need for the stabilization of large biophase loadings in the reactor, coupled with hindrance of bacterial growth under anaerobic conditions [26, 29, 41], suggest that the confinement of the biophase within the reactor is a necessary prerequisite to make the process effective.

4 The Biofilm Reactor as a Tool for Process Intensification

Among the wide choice of reactor designs, the biofilm reactor is one of the best suited for azo-dye conversion as it meets two important process requisites. The first is related to the hindered growth feature of bacterial metabolism under anaerobic conditions. The second is related to the necessity to increase cell densities (see previous section) with respect to those commonly harvested in liquid broths [55, 56]. Except for bacteria that forms aggregates spontaneously, immobilization of cells on granular carriers and membrane reactor technology are the two common pathways to achieve high-density confined cell cultures in either discontinuous or flow reactors.

The role of bacterial biofilm in anaerobic azo-dye conversion has been addressed by several researchers, starting from the pioneering contribution by Jiang and Bishop [3]. Zhang et al. [5] characterized biofilm activity on an aerobic rotating drum reactor (Fig. 2b). They showed that azo-dye conversion still proceeds under aerobic conditions since the progressive uptake of oxygen moving towards the inner region of the biofilm makes local microaerophilic conditions.

On the other hand, cell immobilization on carriers definitively improves bioreactor efficiency. Cell aggregation in a biofilm structure increases process stability and tolerance to shock loadings. A proper selection of operating conditions allows to stabilize a large biophase concentration and to generate smaller amounts of biological sludge.

The choice of solid carriers spans a wide spectrum (Table 1): from materials most suitable for research purposes (sintered glass beads, laterite stone deposited on a gramophone disk) to industrial materials (pumice, activated carbon, etc.). Key properties that affect the performance of the carrier are porosity (from impervious to controlled-size pores), composition (from ceramics to activated carbon), and hydrophilic behavior. It is difficult to perform a direct comparison of different carriers. Colonization and biofilm growth depend strongly on the nature of bacteria and on their intrinsic propensity to adhere on hydrophilic vs. hydrophobic surfaces.

The activated carbon is often chosen as granular carrier and is referred to as granular activated carbon (GAC) or biological activated carbon (BAC). This support is characterized by a strong propensity to adsorb the dyes and is a good candidate for biofilm formation, because of its natural organic matrix. The marked propensity to adsorb dyes contributes to increase the tolerance to shock loadings. In fact, under transient conditions the free surface of the AC acts as a temporary buffer for the dye, which is eventually released as dye disappears.

Table 1 reports a wide spectrum of typologies of biofilm reactor: upflow anaerobic sludge bed (UASB), fluidized bed, airlift, fixed bed with and without recycle, mechanically agitated vessel, rotating drum and rotating biological contactor. Each reactor is characterized by positive features and drawbacks.

The fixed bed reactors may be easily operated. However, clogging phenomena may arise during processes characterized by bacterial growth associated with azo-dye conversion. The reactors may be operated either as PFR or as CSTR. The latter configuration may be realized by the recirculation of the liquid at large recycle ratio R [30].

Fluidized beds, both in the conventional and in the airlift configurations, require more careful operation. Proper selection of the operating conditions makes it possible to control biofilm-growth while preventing reactor clogging. Typically, the reactor is operated as a CSTR by establishing large recycle of the liquid stream [36, 37].

5 Reactor Modeling for Azo-Dye Conversion

Reactor modeling of azo-dye conversion requires the following aspects to be thoroughly represented:

- Reactor hydrodynamics
- Mass transport phenomena between phases present in the bioreactor (gas, liquid and solids)
- Kinetics of the main conversion processes (biophase growth, azo-dye conversion, etc.)

Sub-models depend on the reactor typology and on the biophase state. The broad spectrum of bioreactor types, and associated hydrodynamics, and the different

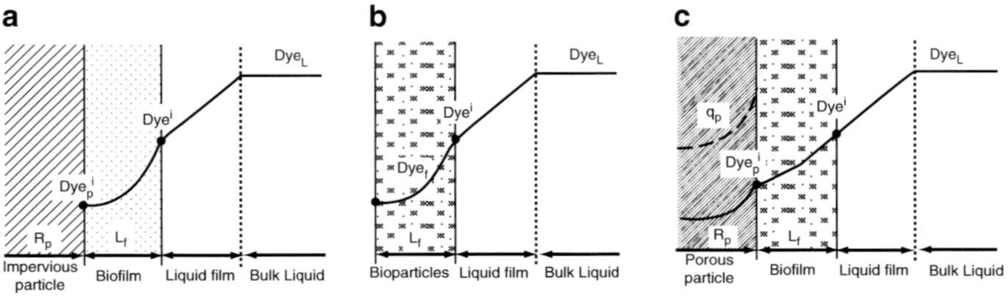

Fig. 4 Profiles of a species diffusing from the bulk liquid towards segregated biophase. (**a**) Biofilm on an impervious particle. (**b**) Aggregated cells or entrapped cells beads; (**c**) Biofilm on a porous particle. q_P: dye concentration on the solid phase

aggregation states of the biophase (free cells, biofilm, entrapped cells, sludge) would justify many different case-tailored models. On the contrary, very few modeling studies can be found in the literature.

The available models mostly refer to ideal reactors, STR, CSTR, continuous PFR. The extension of these models to real reactors should take into account the hydrodynamics of the vessel, expressed in terms of residence time distribution and mixing state. The deviation of the real behavior from the ideal reactors may strongly affect the performance of the process. Liquid bypass – which is likely to occur in fluidized beds or unevenly packed beds – and reactor dead zones – due to local clogging or non-uniform liquid distribution – may be responsible for the drastic reduction of the expected conversion. The reader may refer to chemical reactor engineering textbooks [51, 57] for additional details.

Mass transport phenomena become relevant to the reactor performance as soon as segregated biophases (aggregated, entrapped, biofilm) are used. Figure 4 reports the qualitative patterns of concentration profiles of a compound diffusing from the liquid bulk towards the biophase while bioconversion is active. The figure refers to three possible scenarios. The first (a) regards biofilm growth on impervious particles, the seconds (b) regards cellular aggregated and entrapped cells and the third (c) regards biofilm growth on porous particles. The decrease in the concentration moving from the liquid bulk towards the center of the segregated biophase is governed by the competition between transport phenomena and bioconversion. The latter acts as a sink for the substrates within the segregated biophase. The main difference between cases "a" and "b" is that the profile extends up to the particle center for case "b," provided that the aggregate or entrapment beads behave homogenously. The profile in case "c" shows a discontinuity at the surface of the porous particle. This feature reflects the fact that the biofilm structures within the particles may be different from those formed around the particles. Moreover, the decrease in the dye concentration may be ascribed to biofilm conversion and adsorption. The latter becomes relevant during unsteady state operations.

The conversion products follow profiles that mirror those presented in Fig. 4. A source region of products is localized within the segregated biophase.

Whatever the typology of immobilized biophase, kinetics assessment and modeling studies should not neglect the relevance of the profiles reported in Fig. 4. In agreement with Bailey and Ollis [51], the non uniform profile of the concentrations of azo-dye and of the products may be expressed in terms of the effectiveness factor of the immobilized biophase: the ratio of actual reaction rate to the reaction rate without diffusion limitation.

Chen et al. [54] have reported a model for the assessment of the combined effects of the intrinsic reaction kinetics and dye diffusion into phosphorylated polyvinyl alcohol (PVA) gel beads. The analysis of the experimental data in terms of biofilm effectiveness factor highlighted the relevance of intraparticle diffusion to the effective azo-dye conversion rate. On the basis of these results, they have identified the optimal conditions for the gel bead diameter and PVA composition to limit diffusion resistance.

The knowledge of the kinetics of the dye conversion process and of all phenomena involved in the bacterial life-cycle and maintainance (growth, death, respiration, etc.) in the planktonic and segregated states are a prereqisite to bioreactor design and operation. The first issue has been addressed in the previous section. The relevance of microrganism growth, death and respiration is often neglected in most modeling studies and analysis of experimental data.

The role of cell respiration has been taken into account to interpret the azo-dye conversion by particle-supported biofilm under aerobic conditions [5, 24]. The rapid depletion of oxygen expected/measured as one moves inside the biofilm promotes the establishment of the anoxic conditions needed for azo-dye conversion.

The kinetics of cell growth/death under free and/or immobilized states assume a relevant role in the assessment of the amount of biophase present in the reactor. Obviously, the kinetics depends strongly on the carbon/energy source available in wastewaters or purposely added. With the exception of consortia collected from anaerobic digesters, single strain cultures used in azo-dye conversion are characterized by hindered growth under anaerobic conditions [26, 29, 41]. For these biosystems, the duration of the anaerobic stage must be carefully monitored to preserve cell viability.

Modeling of biofilm reactors should take into account the dynamic equilibrium between competing processes of biofilm growth and detachment. Several phenomena contribute to biofilm detachment – shear-induced "erosion," removal of large patches of biofilm ("sloughing"), "abrasion" – and that their rate depends on the operating conditions. The relevance of biofilm detachment is emphasized in the anaerobic stage when biophase growth is hindered. Russo et al. [58] have shown that careful operation of the reactor must be ensured to control biofilm detachment and to preserve the biophase concentration, which dictates azo-dye conversion. The occurrence of biofilm detachment may emphasize the contribution of free cells even under operating conditions that would be classified as "wash-out" operation. In fact, the suspended biophase produced by biofilm detachment may stabilize a loading of free microrganisms that may be comparable with the immobilized/entrapped phase.

Models available in literature refer to reactors operated/assumed as a CSTR [45, 59] or as a PFR with recycle [36].

The models proposed by Wu et al. [36] and by Lin and Leu [45] refer to continuous conversion processes by immobilized bacteria: the first to a fixed mixed culture entrapped into PVA beads operated in a fluidized bed, and the second to BAC of *P. luteola* operated in a packed bed. Results of these models highlight the role of mass transport phenomena and biophase granule size on reactor performance.

6 A Selected Case Study: *Pseudomonas* Biofilm Reactor for the Conversion of Acid Orange 7

6.1 Experimental

Lodato et al. [41] reported that free cells of *Pseudomonas* sp. OX1 successfully converted Acid Orange 7 under anaerobic conditions. They characterized the conversion process in terms of kinetics (see Table 2) and of the maximum extent of the decolorization (the "decolorization quotient," see Fig. 1). The required sequence of an aerobic stage to promote growth/maintenance of the microorganism and an anaerobic stage to favor cleavage of the azo-bond and decolorization has been established by means of a cyclic process consisting of alternating aerobic–anaerobic phases.

The bioconversion process of Acid Orange 7 will be hereby analyzed. This is an incremental study with respect to that due to Lodato et al. [41], based on the operation of an airlift reactor with cells of *Pseudomonas* sp. OX1 immobilized on natural pumice (density = 1,000 kg/m^3; particle size = 800–1,000 μm). Details regarding the strain, medium, culture growth and main diagnostics of the liquid phase are reported by Lodato et al. [41]. Elemental analysis of dry biomass was obtained by a C/H/N 600 LECO analyzer.

Figure 5 shows a sketch of the experimental apparatus. It consists of a bench scale internal loop airlift, gas and liquid flow control units and a gas humidifier.

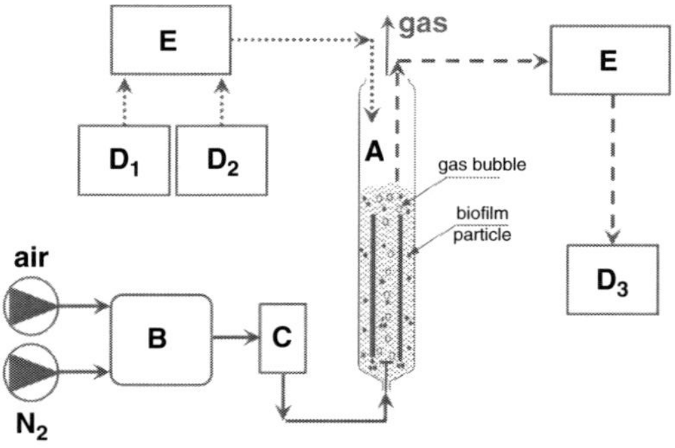

Fig. 5 Experimental apparatus: (A) airlift bioreactor; (B) gas flow control unit; (C) humidifier; (D_1) medium tank; (D_2) dye solution tank; (D_3) wastewater tank; (E) liquid flow control unit

Geometric details of the reactor are reported by [41]. The volume of the liquid phase in the internal loop airlift, hence the reaction volume V, could be changed by varying the level of an overflow duct.

Tests were carried out at 25°C and at initial pH 6.9. Cultures in the liquid medium were incubated in 50 mL Falcon tubes, continuously shaked at 220 rpm. Each culture contained a fresh *Pseudomonas* sp. OX1 colony in 10 mL of medium. The airlift with 10 g of pumice was sterilized at 121°C for 30 min and then housed in a sterile room. One-day culture was transferred to the reactor and, after a batch phase, liquid medium with phenol as the only carbon source was continuously fed. The reactor volume V was fixed at 0.13 L. Aerobic conditions were established sparging technical air. Under these conditions microorganism started to grow immobilized on the solid's support. When immobilized biomass approached steady state, cyclic operation of the airlift was started by alternating aerobic/anaerobic conditions.

- *Aerobic phase*. Technical air and liquid medium were continuously fed to the airlift during the aerobic phase. Gas flow rate was set at 5 nL/h corresponding to 0.64 vvm. The feeding rate of the phenol-bearing (200 mg/L) stream was set at 20 mL/h, that is $D = 0.15\ h^{-1}$. The dilution rate was set at a value larger than the maximum grow rate (wash-out conditions with respect to free cells), $0.14\ h^{-1}$ [60].
- *Anaerobic phase*. Nitrogen was sparged at 5 nL/h and the liquid feeding was stopped. The concentration of acid orange 7 at the beginning of the anaerobic phase was set at the pre-fixed value by injecting concentrated dye solution into the reactor. The reactor was operated under batch conditions with respect to the liquid phase.

During the run a sampling port was used to perform aseptic sampling of the culture. Sampling enabled to measure cell, dye and carbon source concentrations during the test.

Bioparticles were periodically sampled and their carbon content measured. The mass of dry biofilm-cells on solid carriers was estimated assuming the following: (1) the carbon content is related only to the contribution of biofilm-cells, neglecting the contribution of extra cellular matter; (2) the carbon mass fraction of immobilized cells equals the value estimated for free cells, namely 0.44 [60].

Figure 6 reports selected representative data regarding the decolorization process. Aerobic/anaerobic cycling extended up to 2 months. Data refer to the concentrations of phenol – the carbon source – and acid orange 7. Vertical lines mark the times when switching between aerobic (A) and anaerobic (AN) conditions occurred. The concentration of azo-dye at the beginning of each anaerobic phase was fixed at value ranging between 35 and 140 mg/L. The biofilm concentration, assessed as carbon equivalent, ranged between 0.8 and 1.2 g_{DM}/L.

Results suggest that the biofilm was stable over 2 months. The microorganism appeared vital and active even after repeated cyclic exposure to anaerobic conditions. In fact, phenol uptake measured was always recorded as soon as aerobic conditions were established. A slight increase of the phenol concentration was

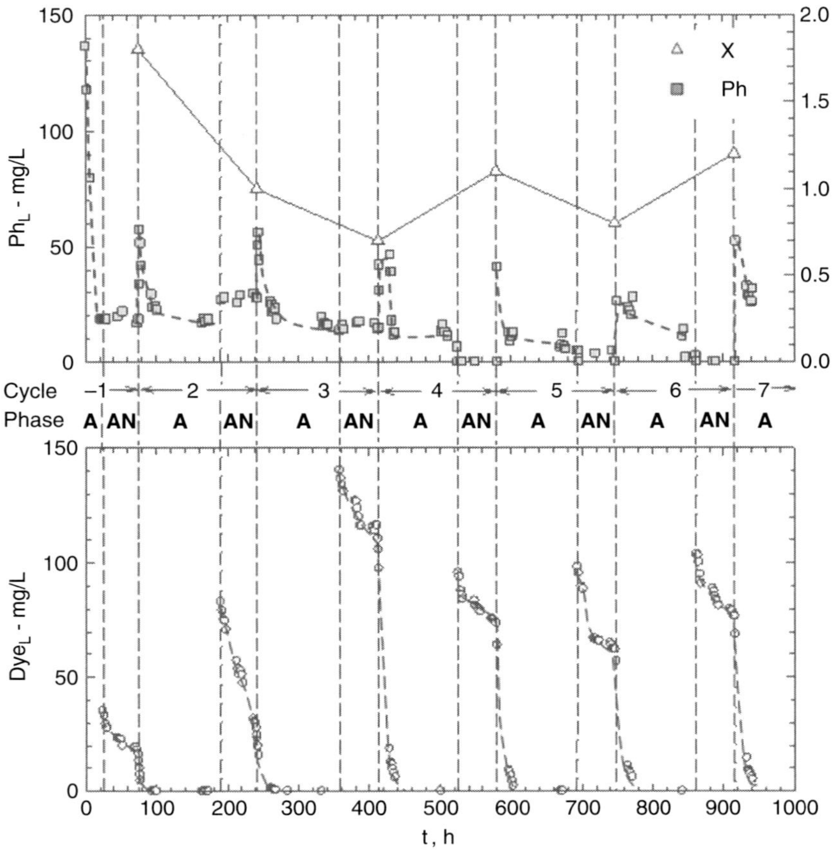

Fig. 6 Acid orange 7 and phenol concentration in the internal loop airlift reactor operated with *Pseudomonas* sp. *OX1* biofilm on natural pumice. (A) Aerobic phase. Gas: air. Liquid: continuous feeding of phenol supplemented synthetic medium. (AN) Anaerobic phase. Gas: nitrogen. Liquid: batch conditions, dye supplemented medium

typically observed at the beginning of each aerobic phase. The maximum could be interpreted by taking into account the competition that establishes at the beginning of the aerobic phase between the phenol convective flow and the likely "relaxation" time of the microorganism to re-activate the aerobic metabolic pathway. Phenol concentration approached the steady state value in over about 2 days. Dye conversion occurs only during the anaerobic phase. Assuming that the biofilm load was about constant during the process, the slope of the Dye_L vs. t at the beginning of the run increased with the dye concentration set at the beginning of the anaerobic phase. In agreement with (9), the decolorization rate increased with dye concentration. The value of the specific dye conversion rate was nearly equal to that assessed for *Pseudomonas* sp. OX1 free cells by [41]. The sharp decrease of dye concentration during the aerobic phase was due to the reactor wash-out by the phenol-bearing stream fed to the reactor.

6.2 Dynamic Modeling

A dynamic model has been developed to simulate the behavior of a *Pseudomonas* sp. OX1 biofilm reactor for phenol and azo-dye conversion during the aerobic–anaerobic cyclic operation. Phenol and oxygen were considered as the limiting substrates for growth kinetics.

The model relies on the following assumptions:

- The reactor flow pattern is that of an internal loop airlift with pneumatic mixing of both the liquid and the solid phases [61], the latter consisting in biofilm supported by granular solids. The reactor was assumed uniformly mixed.
- The conversion process within the biofilm is described by a substrate diffusion-reaction model.
- The continuous exchange of cells between the biofilm (X) and the liquid phase (X_L) was described by means of a combined attachment/detachment mechanism. The net rate of detachment balances biofilm growth under steady state conditions.

Model computations were directed to simulate dye conversion and biophase growth under alternated aerobic/anaerobic conditions.

Aerobic: The growth kinetics was described by an interacting, balanced and unstructured model characterized by phenol inhibition and oxygen limitation according to a double limiting kinetics [60, 62].

$$\frac{1}{Y_{X/\mathrm{Ph}}}\mathrm{Ph}+\frac{1}{Y_{X/O_2}}O_2 \rightarrow X \qquad \mu=\mu^{M}\frac{\mathrm{Ph}}{K_{\mathrm{Ph}}+\mathrm{Ph}+\mathrm{Ph}^2/K_I}\frac{O_2}{K_{O_2}+O_2}. \tag{8}$$

Anaerobic: Azo-dye is converted batchwise. Dye conversion was described in agreement with [41]. Inhibition by oxygen was taken into account according to a non-competitive mechanism.

$$r_{\mathrm{Dye}}=\frac{r_{\mathrm{Dye}}^{\max}\times \mathrm{Dye}}{K_{\mathrm{Dye}}+\mathrm{Dye}}\frac{K_{O_2}^{I}}{K_{O_2}^{I}+O_2}X. \tag{9}$$

Moreover, the following assumptions were made:

- External mass transfer resistance was neglected, as reported by [63]: in biofilm reactors with granular particles (fluidized bed, airlift) the Biot number was generally larger than 100.

Table 3 Model equations (mass balance on liquid phase and on biofilm) and parameter values

Liquid phase		
Phenol	$\frac{d\mathrm{Ph_L}}{dt}=D\left(\mathrm{Ph_L^{IN}}-\mathrm{Ph_L}\right)-\frac{\mu^M}{Y_{X/Ph}}\frac{\mathrm{Ph_L}}{K_{Ph}+\mathrm{Ph_L}+\mathrm{Ph_L}^2/K_I}\frac{\mathrm{O_{2L}}}{K_{O_2}+\mathrm{O_{2L}}}\left(X_L+\frac{\eta_X 3L_f}{R_P X\varepsilon_S}\right)$	(T.3.1)
Oxygen	$\frac{d\mathrm{O_{2L}}}{dt}=\left(D+K_L{}^{O_2}a_L\right)\left(\mathrm{O_{2L}^{Eq}}-\mathrm{O_{2L}}\right)-\frac{\mu^M}{Y_{X/O_2}}\frac{\mathrm{Ph_L}}{K_{Ph}+\mathrm{Ph_L}+\mathrm{Ph_L}^2/K_I}\frac{\mathrm{O_{2L}}}{K_{O_2}+\mathrm{O_{2L}}}\left(X_L+\eta_X 3L_f/R_P X\varepsilon_S\right)$	(T.3.2)
Dye	$\frac{d\mathrm{Dye_L}}{dt}=-\mathrm{D}\cdot\mathrm{Dye_L}-r_{dye}^M\frac{\mathrm{Dye_L}}{K_{Dye}+\mathrm{Dye_L}}\frac{K_{O_2}^I}{K_{O_2}^I+\mathrm{O_{2L}}}\left(X_L+\frac{\eta_X 3L_f}{R_P X\varepsilon_S}\right)$	(T.3.3)
Immobilized biomass	$3\varepsilon_S X\frac{dL_f/R_P}{dt}V_L=K_A X_L V_L-K_D 3L_f/R_P\varepsilon_S X+\eta_X\cdot\mu^M\frac{\mathrm{Ph_L}}{K_{Ph}+\mathrm{Ph_L}+\mathrm{Ph_L}^2/K_I}\frac{\mathrm{O_{2L}}}{K_{O_2}+\mathrm{O_{2L}}}3L_f/R_P\varepsilon_S X$	(T.3.4)
Free biomass	$\frac{dX_L}{dt}=-DX_L-K_A X_L+K_D 3L_f/R_P\varepsilon_S X+\frac{\mathrm{Ph_L}}{K_{Ph}+\mathrm{Ph_L}+\mathrm{Ph_L}^2/K_I}\frac{\mathrm{O_{2L}}}{K_{O_2}+\mathrm{O_{2L}}}X_L$	(T.3.5)
Biofilm phase		
Phenol	$\frac{\partial \mathrm{Ph}_r}{\partial t}=\frac{1}{r^2}\frac{\partial}{\partial r}r^2\mathcal{D}_B^{Ph}\frac{\partial \mathrm{Ph}_r}{\partial r}-\frac{\mu^M}{Y_{X/Ph}}\frac{\mathrm{Ph}_r}{K_{Ph}+\mathrm{Ph}_r+\mathrm{Ph}_r^2/K_I}\frac{\mathrm{O}_{2r}}{K_{O_2}+\mathrm{O}_{2r}}$; $\mathrm{Ph}_{r=L_f}=\mathrm{Ph_L}$; $\frac{\partial \mathrm{Ph}}{\partial r}_{r=0}=0$	(T.3.6a-b-c)
Oxygen	$\frac{\partial \mathrm{O}_{2r}}{\partial t}=\frac{1}{r^2}\frac{\partial}{\partial r}r^2\mathcal{D}_{\mathcal{B}}^{O_2}\frac{\partial \mathrm{O}_{2r}}{\partial r}-\frac{\mu^M}{Y_{X/O_2}}\frac{\mathrm{Ph}_r}{K_{Ph}+\mathrm{Ph}_r+\mathrm{Ph}_r^2/K_I}\frac{\mathrm{O}_{2r}}{K_{O_2}^*+\mathrm{O}_{2r}}$; $\mathrm{O}_{2r=L_f}=\mathrm{O_{2L}}$; $\frac{\partial \mathrm{O_2}}{\partial r}_{r=0}=0$	(T.3.7a-b-c)
Dye	$\frac{\partial \mathrm{Dye}_r}{\partial t}=\frac{1}{r^2}\frac{\partial}{\partial r}r^2\mathcal{D}_B^{Dye}\frac{\partial \mathrm{Dye}_r}{\partial r}-r_{dye}^M\frac{\mathrm{Dye}_r}{K_{Dye}+\mathrm{Dye}_r}\frac{K_{O_2}^I}{K_{O_2}^I+\mathrm{O}_{2r}}$; $\mathrm{Dye}_{r=L_f}=\mathrm{Dye_L}$; $\frac{\partial \mathrm{Dye}}{\partial r}_{r=0}=0$	(T.3.8a-b-c)

- The biofilm thickness (L_f) and density ($X = 50$ g/L) were assumed uniform and the biofilm treated as a continuum. A substrate diffusion-reaction model assuming spherical particle was used. Diffusion coefficient of phenol and oxygen in the biofilm were assessed according to Fan et al. [64]:

$$\frac{\mathcal{D}_B}{\mathcal{D}_L} = 1 - \frac{0.43X^{0.92}}{11.19 + 0.27X^{0.99}}. \qquad (10)$$

- Adhesion of suspended cell to the bioparticles was modeled through a first-order kinetics with respect to suspended biomass [58, 65]. Gjaltema et al. [66, 67] reported that specific detachment rate in airlift was mainly due to the particle-to-particle collisions. Accordingly, it was assumed that the detachment rate was proportional to the immobilized biomass concentration.

$$r_{adh} = K_A X_L \qquad r_{det} = \frac{K_D 3L_f}{R_P \varepsilon_S X}. \qquad (11)$$

Table 3 reports the balance equations on phenol, oxygen and dye in the liquid phase and in the biofilm, together with equations expressing the suspended biomass concentration and the biofilm thickness. The dilution rate was set to zero during the anaerobic stage. The quasi-steady state approximation was made in developing the balance on biofilm, justified by the condition that diffusion across the biofilm is much faster than growth kinetics. Simulations were carried out assuming input values reported in Table 4.

Figure 7 shows phenol, dye, oxygen and suspended biomass concentrations and biofilm thickness as a function of time.

- *Aerobic phase*: Steady state values of phenol concentration (40 mg/L) and biofilm thickness (170 μm) were approached after a 5 h transient period, which reproduces fairly well the experimental dynamical patterns reported in Fig. 6. However, biomass was present also in the liquid phase as a consequence of biofilm detachment.

Table 4 Parameters and operating conditions that are input to the model

$\mu^M = 0.71$ h^{-1} [a]		$K_{Ph} = 310$ mg/L [a]	$K_I = 130$ mg/L [a]
$K_{O2} = 0.1$ mg/L		$K_{Dye} = 290$ mg/L [b]	$K^I_{O_2} = 0.1$ mg/L [b]
$Y_{X/Ph} = 0.74$ [a]		$Y_{X/O2} = 0.17$	
$X = 50$ mg/L		$\varepsilon_S = 10\%$	$R_P = 0.15$ mm
$Ph_L^{IN} = 180$ mg/L		$Dye_L(t = t_0^{AN}) = 120$ mg/L	$K_A = 0.0012$ h^{-1}
$K_D = 0.01$ h^{-1}		$K_L a_L = 100$ h^{-1}	
		Aerobic phase	**Anaerobic phase**
	O_{2L}^{Eq}	7.8 mg/L	0 mg/L
	D	0.17 h^{-1}	0 h^{-1}

[a] [60], [b] [41]

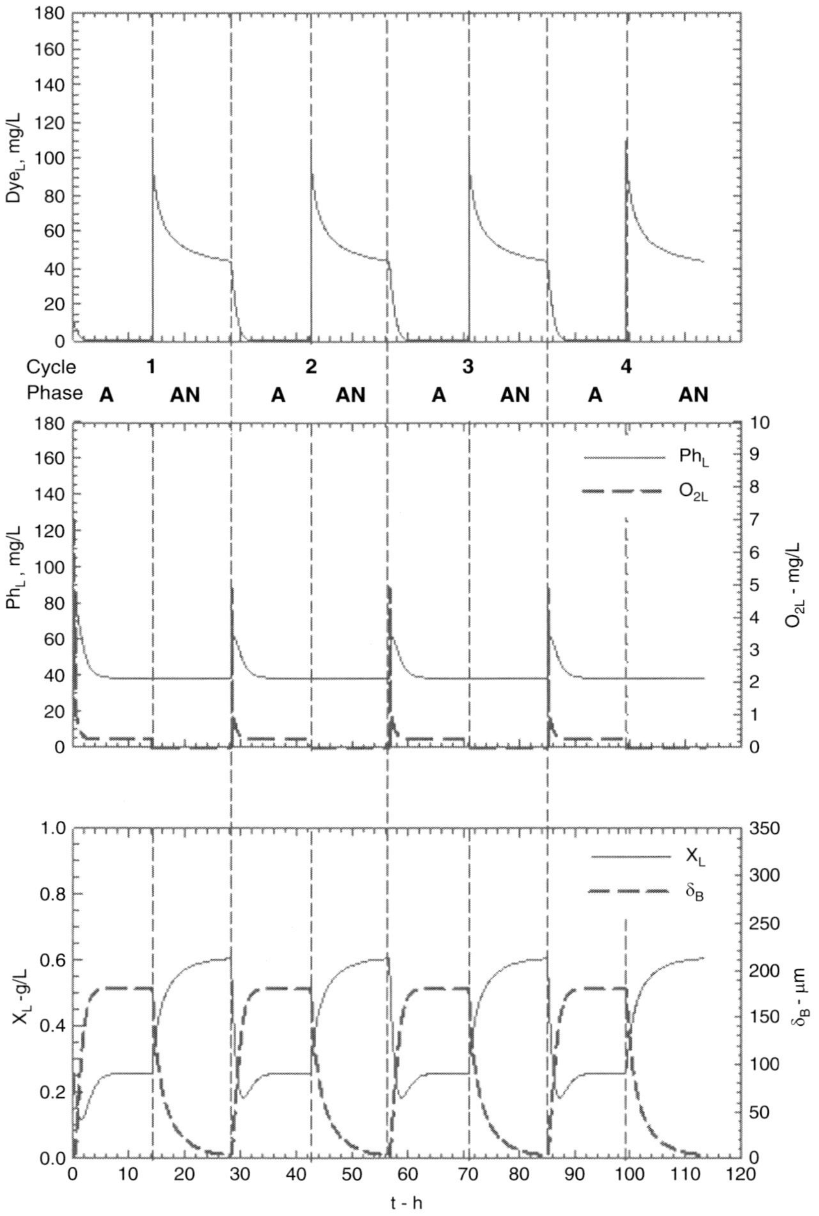

Fig. 7 Numerical simulation of aerobic–anaerobic sequential process

- *Anaerobic phase*: The dissolved oxygen quickly vanishes due to nitrogen sparging. Eventually, dye concentration decreases following a pattern that closely reproduces those observed in experiments. It is worth noting that the thickness of the biofilm decreases during this stage. In fact, detachment mechanism is not balanced by growth, which is absent in anaerobic condition.

Model computations suggest that cyclic operation is able to regenerate during the aerobic stage the reducing potential for dye conversion during the anaerobic stage.

7 Concluding Remarks and Research Needs

Though extensive research has been recently reported in the open literature concerning biodegradation of azo-dyes, there is still a need for additional research. The close link between dye structure and reaction pathways and rates makes it difficult to extrapolate results obtained with one dye to that with others, even belonging to the same class. Dye-specific assessment of process rates and yields is therefore required. Also, the degradation potential of several bacteria or consortia active toward xenobiotics in dye degradation is far from being fully assessed.

Coupling of aerobic–anaerobic stages has proven to be effective in promoting biomass growth and maintainance, build-up of reducing compounds required for dye degradation, conversion of dyes and further degradation of intermediates from primary dye bioconversion. Better understanding of the metabolic pathways relevant to alternated aerobic–anaerobic operation could greatly contribute to process optimization.

Biomass containment in continuously operated bioreactors is an essential prerequisite for the feasibility of practical industrial-scale dye biodegradation. Biofilm airlift reactors have demonstrated excellent performance for their ability to control mixing, interphase mass transfer and biofilm detachment rate. Further studies are required to further exploit the potential of this type of reactors with either aggregated cells or biofilm supported on granular carriers.

Nomenclature

D	Dilution rate	h^{-1}
$\mathcal{D}_J$	Diffusivity in J phase	cm^2/s
Dye	Dye concentration	mg/L
ε_S	Solids holdup	–
HRT	Hydraulic residence time	h
K_A	Adhesion coefficient	h^{-1}
K_D	Detachment coefficient	h^{-1}
K_{Dye}	Michaelis–Menten dye coefficient	mg/L
K_I	Inhibition phenol coefficient	mg/L
K_{O2}	Oxygen coefficient	mg/L
$K^I_{O_2}$	Inhibition oxygen coefficient	mg/L
K_{Ph}	Phenol coefficient	mg/L
$K_L^{O2}a_L$	Oxygen mass transfer coefficient	h^{-1}
μ	Specific growth rate	h^{-1}
L_f	Biofilm thickness	mm

O_2	Dissolved oxygen concentration	mg/L
Ph	Phenol concentration	mg/L
Q	Liquid flow rate	L/h
R	Recycling ratio	–
R_B	Biofilm radius	mm
R_P	Particle radius	mm
r_{Dye}	Dye conversion rate	mg/(L h)
S	Substrate concentration	g/L
t	Time	s
T	Temperature	°C
V	Liquid volume	L
X	Biomass concentration	g/L
Y	Yield coefficient	g/g
W	Mass flow rate	g/h
Subscripts and Superscripts		
0	Initial condition	
adh	Adhesion	
C	Carbon	
det	Detachment	
i	Interface	
L	Liquid phase	
P	Particle	

Acknowledgments The support of Dr. Antonello Lodato and Dr. Fabiana Alfieri in experimental investigation is gratefully acknowledged.

References

1. Pearce CI, Lloyd JR, Guthrie JT (2003) The removal of colour from textile wastewater using whole bacterial cells: a review. Dyes Pigm 58:179–196
2. van der Zee FP, Villaverde S (2005) Combined anaerobic–aerobic treatment of azo dyes – a short review of bioreactor studies. Water Res 39:1425–1440
3. Jiang H, Bishop PL (1994) Aerobic biodegradation of azo dyes in biofilms. Water Sci Technol 29:525–530
4. Carliell CM, Barclay SJ, Naidoo N et al (1995) Microbial decolorization of a reactive azo dye under anaerobic conditions. Water SA 21:61–69
5. Zhang TC, Fu YC, Bishop PL et al (1995) Transport and biodegradation of toxic organics in biofilms. J Hazard Mater 41:267–285
6. Razo-Flores E, Luijten M, Donlon BA et al (1997) Complete biodegradation of the azo dye azodisalicylate under anaerobic conditions. Environ Sci Technol 31:2098–2103
7. Chang JS, Lin YC (2000) Fed-batch bioreactor strategies for microbial decolorization of azo dye using *Pseudomonas Luteola* strain. Biotechnol Progr 16:979–985
8. Cruz A, Buitron G (2000) Biodegradation of disperse blue 79 using sequenced anaerobic/aerobic biofilters. Water Sci Technol 44:159–166
9. O'Neill C, Hawkes FR, Hawkes DL et al (2000) Anaerobic–aerobic biotreatment of simulated textile effluent containing varied ratios of starch and azo dye. Water Res 34:2355–2361
10. Bras R, Ferra MIA, Pinheiro HM et al (2001) Batch tests for assessing decolorization of azo dyes by methanogenic and mixed cultures. J Biotechnol 89:155–162

11. Chang JS, Chou C, Chen SY (2001) Decolorization of azo dyes with immobilized *Pseudomonas luteola*. Process Biochem 36:757–763
12. Chang JS, Chou C, Lin YC et al (2001) Kinetic characteristics of bacterial azo-dye decolorization by *Pseudomonas Luteola*. Water Res 35:2841–2850
13. van der Zee FP, Lettinga G, Field JA (2001) Azo dye decolorisation by anaerobic granular sludge. Chemosphere 44:1169–1176
14. Yu J, Wang X, Yue PL (2001) Optimal decolorization and kinetic modeling of synthetic dyes by *pseudomonas* strains. Water Res 35:3579–3586
15. Coughlin MF, Kinkle BK, Bishop PL (2002) Degradation of acid orange 7 in an aerobic biofilm. Chemosphere 46:11–9
16. Yoo ES (2002) Kinetic of chemical decolorization of the azo-dye C.I. reactive orange 96 by sulfide. Chemosphere 47:925–931
17. Abraham TE, Senan RC, Shaffiqu TS et al (2003) Bioremediation of textile azo dyes by an aerobic bacterial consortium using a rotating biological contactor. Biotechnol Progr 19:1372–1376
18. Chen K-C, Wu J-Y, Huang CC (2003) Decolorization of azo-dye using PVA-immobilized microorganism. J Biotechnol 101:241–252
19. Coughlin MF, Kinkle BK, Bishop PL (2003) High performance degradation of azo dye Acid Orange 7 and sulfanilic acid in a laboratory scale reactor after seeding with cultured bacterial strains. Water Res 37:2757–2763
20. Isik M, Sponza DT (2003) Effect of oxygen on decolorization of azo dyes by *Escherichia coli* and *Pseudomonas* sp. and fate of aromatic amines. Process Biochem 38:1183–1192
21. Senan RC, Shaffiqu TS, Roy JJ et al (2003) Aerobic degradation of a mixture of azo dyes in a packed bed reactor having bacteria-coated laterite pebbles. Biotechnol Progr 19:647–651
22. Buitron G, Quezada M, Moreno G (2004) Aerobic degradation of the azo dye Acid Red 151 in a sequencing batch biofilter. Biores Technol 92:143–149
23. Isik M, Sponza DT (2004) A batch kinetic study on decolorization and inhibition of Reactive Black 5 and Direct Brown 2 in an anaerobic mixed culture. Chemosphere 55:119–128
24. Li J, Bishop PL (2004) Adsorption and biodegradation of azo dye in biofilm processes. Water Sci Technol 49:237–245
25. Moutaouakkil A, Zeroual Y, Dzayri FZ et al (2004) Decolorization of azo dyes with Enterobacter agglomerans immobilized in different supports by using fluidized bed bioreactor. Curr Microbiol 48:124–129
26. Yeh MS, Chang JS (2004) Bacterial decolorization of an azo dye with a natural isolate of *Pseudomonas luteola* and genetically modified *Escherichia coli*. J Chem Technol Biotechnol 79:1354–1360
27. Beydilli MI, Pavlostathis SG (2005) Decolorization kinetics of the azo dye Reactive Red 2 under methanogenic conditions. Effect of long-term culture acclimation. Biodegradation 16:135–146
28. Bras R, Gomes A, Ferra MIA et al (2005) Monoazo and diazo dye decolorization studies in a methanogenic UASB reactor. J Biotechnol 115:57–66
29. Chen BY, Chen SY, Chang JS (2005) Immobilized cell fixed-bed bioreactor for wastewater decolorization. Process Biochem 40:3434–3440
30. Georgiou D, Hatiras J, Aivasidis A (2005) Microbial immobilization in a two-stage fixed-bed-reactor pilot plant for on-site anaerobic decolorization of textile wastewater. Enzyme Microb Technol 37:597–605
31. Maas R, Chaudhari S (2005) Adsorption and biological decolourization of azo dye Reactive Red 2 in semicontinuous anaerobic reactors. Process Biochem 40:699–705
32. Ong SA, Toorisaka E, Hirata M et al (2005) Treatment of azo dye orange II in a sequential anaerobic and aerobic-sequencing batch reactor system. Environ Chem Lett 2:203–207
33. Ong SA, Toorisaka E, Hirata M et al (2005) Decolorization of azo dye (Orange II) in a sequential UASB-SBR system. Sep Pur Technol 42:297–302

34. Ong SA, Toorisaka E, Hirata M et al (2005) Treatment of azo dye Orange II in aerobic and anaerobic-SBR systems. Process Biochem 40:2907–2914
35. Sandhya S, Padmavathy S, Swaminathan K et al (2005) Microaerophilic-aerobic sequential batch reactor for treatment of azo dyes containing simulated wastewater. Process Biochem 40:885–890
36. Wu JY, Hwang SCJ, Chen CT et al (2005) Decolorization of azo dye in a FBR reactor using immobilized bacteria. Enzyme Microb Technol 37:102–112
37. Georgiou D, Aivasidis A (2006) Decoloration of textile wastewater by means of a fluidized-bed loop reactor and immobilized anaerobic bacteria. J Hazard Mater 135:372–377
38. Isik M, Sponza DT (2006) Biological treatment of acid dyeing wastewater using a sequential anaerobic/aerobic reactor system. Enzyme Microb Technol 38:887–892
39. Khehra MS, Saini HS, Sharma DK (2006) Biodegradation of azo dye C.I. Acid Red 88 by an anoxic–aerobic sequential bioreactor. Dyes Pigm 70:1–7
40. Lourenco ND, Novais JM, Pinheiro HM (2006) Kinetic studies of reactive azo dye decolorization in anaerobic/aerobic sequencing batch reactors. Biotechnol Lett 28:733–739
41. Lodato A, Alfieri F, Olivieri G et al (2007) Azo-dye conversion by means of *Pseudomonas* sp. OX1. Enzyme Microb Technol 41:646–652
42. Mezohegyi G, Kolodkin A, Castro UI et al (2007) Effective anaerobic decolorization of azo dye Acid Orange 7 in continuous upflow packed-bed reactor using biological activated carbon system. Ind Eng Chem Res 46:6788–6792
43. Singh P, Sanghi R, Pandey A (2007) Decolorization and partial degradation of monoazo dyes in sequential fixed-film anaerobic batch reactor (SFABR). Bioresour Technol 98:2053–2056
44. Dafale N, Wate S, Meshram S et al (2008) Kinetic study approach of remazol black-B use for the development of two-stage anoxic-oxic reactor for decolorization/biodegradation of azo dyes by activated bacterial consortium. J Hazard Mater 159:319–328
45. Lin HY, Leu JY (2008) Kinetic of reactive azo-dye decolorization by *Pseudomonas Luteola* in biological actived carbon process. Biochem Eng J 39:457–467
46. Mezohegyi G, Bengoa C, Stuber F et al (2008) Novel bioreactor design for decolourisation of azo dye effluents. Chem Eng J 143:293–298
47. Ong SA, Toorisaka E, Hirata M et al (2008) Combination of adsorption and biodegradation processes for textile effluent treatment using a granular activated carbon-biofilm configured packed column system. J Environ Sci 20:952–956
48. Ong SA, Toorisaka E, Hirata M et al (2008) Granular activated carbon-biofilm configured sequencing batch reactor treatment of C.I. Acid Orange 7. Dyes Pigm 76:142–146
49. Wuhrmann K, Mechsner K, Kappeler T (1980) Investigation of rate-determining factors in the microbial reduction of azo dyes. Eur J Appl Microb Biotech 9:325–338
50. Kalyuzhnyi SV (2001) Environmental biotechnology: a tandem of biocatalytic and engineering approaches. Russ Chem Bull 50:1818–1825
51. Bailey JE, Ollis DF (1986) Biochemical engineering fundamentals. McGraw-Hill, New York
52. Levenspiel O (1999) Chemical reaction engineering, 3rd edn. Wiley, New York, NY
53. Russo ME, Giardina L, Marzocchella A et al (2008) Assessment of anthraquinone-dye conversion by free and immobilized crude laccase mixtures. Enzyme Microb Technol 42:521–530
54. Chen K-C, Wu J-Y, Yang W-B et al (2003) Evaluation of effective diffusion coefficient and intrinsic kinetic parameters on azo dye biodegradation using PVA-immobilized cell beads. Biotechnol Bioeng 83:821–832
55. Nicolella C, van Loosdrecht MCM, Heijnen JJ (2000) Wastewater treatment with particulate biofilm reactors. J Biotechnol 80:1–33
56. Qureshi N, Annous BA, Ezeji TC et al (2005) Biofilm reactors for industrial bioconversion processes: employing potential of enhanced reaction rates. Microb Cell Fact 4(24):1–21
57. Westerterp KR, Van Swaaij WPM, Beenackers AACM (1993) Chemical reactor design and operation. Wiley, New York

58. Russo ME, Maffettone PL, Marzocchella A et al (2008) Bifurcational and dynamical analysis of a continuous biofilm reactor. J Biotechnol 135:295–303
59. Isik M, Sponza DT (2005) Substrate removal kinetics in an upflow anaerobic sludge blanket reactor decolorising simulated textile wastewater. Process Biochem 40:1189–1198
60. Viggiani A, Olivieri G, Siani L et al (2006) An airlift biofilm reactor for the biodegradation of phenol by *Pseudomonas stutzeri* OX1. J Biotechnol 123:464–477
61. Chisti Y (1989) Airlift bioreactors. Elsevier, London
62. Hill GA, Robinson CW (1975) Substrate inhibition kinetics phenol degradation by pseudomonas putida. Biotechnol Bioeng 17:1599–1615
63. Tanyolaç A, Beyenal H (1997) Prediction of average biofilm density and performance of a spherical bioparticle under substrate inhibition. Biotechnol Bioeng 56:319–329
64. Fan LS, Leyva-Ramos R, Wisecarver KD et al (1990) Diffusion of phenol through a biofilm grown on activated carbon particles in a draft-tube three-phase fluidized-bed bioreactor. Biotechnol Bioeng 35:279–286
65. Rijnaarts HHM, Norde W, Bouwer EJ et al (1995) Reversibility and mechanism of bacterial adhesion. Colloids Surf B 4:5–22
66. Gjaltema A, Tijhuis L, van Loosdrecht MCM et al (1995) Detachment of biomass from suspended nongrowing spherical biofilms in airlift reactors. Biotechnol Bioeng 46:258–269
67. Gjaltema A, Vinke JL, van Loosdrecht MCM et al (1997) Abrasion of suspended biofilm pellets in airlift reactors: importance of shape, and particle concentrations. Biotechnol Bioeng 53:88–89
68. Nielsen J, Villadsen J (2002) Bioreaction engineering principles, 2nd edn. Kluwer Academic/ Plenum Publishers, Dordrecht, Netherlands

Treatment of Azo Dye-Containing Wastewater Using Integrated Processes

Xujie Lu and Rongrong Liu

Abstract Azo dyes are the most widely used dyes in textile industry. During the dyeing process, the degree of exhaustion of dyes is never complete, resulting in azo dye-containing effluents. The biodegradation of azo dyes is difficult due to their complex structure and synthetic nature. The removal of azo dyes from industry effluents is desirable not only for aesthetic reasons but also because azo dyes and their breakdown products are toxic to aquatic life and mutagenic to humans. In recent years, application of integrated processes for treatment of azo dye-containing wastewater has received considerable attention in the literatures. This review highlights some of the notable examples in the use of integrated processes for azo dye-containing wastewater treatment and deals with biodegradation mechanism of azo dyes. The review also summarizes and attempts to compare the advantages and disadvantages of integrated processes. It can be found that integrated treatment system seems to be an efficient and promising alternative for the treatment of azo dye-containing wastewater.

Keywords Advanced oxidation–biological processes, Anaerobic–aerobic bacterial process, Azo dyes, Biodegradation, Bioreactor, Decolorization, Integrated processes, Textile wastewater

Contents

X. Lu (✉)
School of Chemical and Environmental Engineering, Jianghan University, Wuhan 430056, China
e-mail: xujie_lu@163.com

R. Liu
College of Environmental Science and Engineering, Donghua University, Shanghai 201620, China

H. Atacag Erkurt (ed.), *Biodegradation of Azo Dyes*,
Hdb Env Chem (2010) 9: 133–155, DOI 10.1007/698_2009_47,

Abbreviations

ADMI	American dye manufacturer institute
AOMBR	Anaerobic–oxic membrane bioreactor
AOPs	Advanced oxidation processes
AR151	Acid Red-151
ASP	Activated sludge process
BOD	Biochemical oxygen demand
COD	Chemical oxygen demand
DO	Dissolved oxygen
EGSB	Expanded granular sludge bed
GAC	Granular activated carbon
HRP	Horseradish peroxidase enzyme
HRT	Hydraulic retention time
MF	Microfiltration
NF	Nanofiltration
RB5	Reactive Black 5
RO	Reverse osmosis
SBR	Sequencing batch reactor
SRT	Solids retention time
TDS	Total dissolved solids
TiO_2	Titan dioxide
TS	Total solids
UASB	Up-flow anaerobic sludge blanket
UF	Ultrafiltration

1 Introduction

Azo dyes are one of the most important types of dyes that are extensively used in textile, leather, food, cosmetics, and paper product industries due to a more economical synthesis, firmness, and greater variety of color compared to natural dyes. The annual world production of azo dyes is estimated to be around one million tons, and more than 2,000 structurally different azo dyes are currently in use [1, 2]. Azo dyes are characterized by the presence of one or more group (–N=N–) bound to large number of aromatic rings such as benzene and naphthalene [3]. The color of azo dyes is due to azo bond and associated chromophores [4, 5]. In general, synthetic azo dyes with substitutions as part of their structure are highly resistant to degradation. Several azo dyes and their reductive metabolism products are toxic. Most of the azo dyes have no carcinogenicity, but the carcinogenicity of an azo dye may be due to the aryl amines derived via the reduction transformation of azo dyes. In addition, dye industry effluent also contains other environmental contaminants. Some of these are additives used in the dyeing process. Thus, dye industry effluent is a significant source of environmental pollution [6, 7].

The removal of dyes from industry effluents is desirable not only for aesthetic reasons but also because azo dyes and their breakdown products are toxic to aquatic life and mutagenic to humans [8, 9]. Without adequate treatment these azo dyes are stable and can remain in the environment for an extended period of time. Consequently, azo dyes have to be removed from wastewaters before discharge.

In recent years, various chemical, physical, and biological treatment methods have been developed for the removal of azo dyes from waters and wastewaters to decrease their impact on the environment. Because of the high cost and disposal problems, many of these methods for treating azo dye-containing wastewater have not been widely applied in the textile industries [10, 11]. A literature survey shows that research has been and continues to be conducted in the areas of integrated processes, including physical, chemical, and biological process, to improve the biodegradation of dyestuffs and minimize the sludge production. Indeed, potential advantages of the strategy of combined chemical and biological processes to treat contaminants in wastewater have been previously suggested [12, 13].

Despite its many potential advantages, and the ever-increasing number of publications, there has never been any attempt to collate all this in a review. Hence, the objectives of the review are the compilation of the currently available literature on the newer achievements in the integrated treatment technologies which have been used to treat azo dye-containing wastewaters, classification and short description of the methods, critical evaluation of the technology processes, and the comparison of their advantages and disadvantages.

2 Combined Physical–Biological Processes

Physical color removal technologies that were reported in the literature include adsorption, chemical precipitation, and membrane separation [14–17]. The high cost and disposal problems have opened the door for further investigation of new techniques. The inability of biological treatment processes in degrading azo dye compounds makes physical treatment a necessary stage prior to biotreatment in order to reduce the burden of the followed processes. Various combined physical–biological processes are applied to treat textile industry effluents to meet regulatory discharge limits.

2.1 Adsorption Mechanisms for Azo Dyes' Removal

In general, the mechanism for azo dyes' removal by adsorption on an adsorbent material may be assumed to involve the following four steps:

1. Bulk diffusion: Migration of azo dyes from the bulk of the solution to the surface of the adsorbent
2. Film diffusion: Diffusion of azo dyes through the boundary layer to the surface of the adsorbent
3. Pore diffusion or intraparticle diffusion: Transport of the azo dyes from the surface to within the pores of the particle
4. Chemical reaction: Adsorption of azo dyes at an active site on the surface of material via ion-exchange, complexation, and/or chelation

And the most important steps are film diffusion, pore diffusion, and chemical reaction. Previous studies showed that amine sites were the main reactive groups for azo dyes, though hydroxyl groups might contribute to adsorption. It is now recognized that chemisorption (ion-exchange, electrostatic attractions) is the most prevalent mechanism, with pH value as the main factor affecting adsorption. Chemisorption, a strong type of adsorption in which molecules are not exchanged but electrons may be exchanged, is commonly cited as the main mechanism for the adsorption of anionic dyes in acidic conditions. According to Fig. 1, the mechanism is briefly described: in the presence of H^+, the amino groups of chitosan become protonated; also, in aqueous solution, the anionic dye is first dissolved and the sulfonate groups in the case of acid or reactive dyes dissociate and are converted to anionic dye ions; the adsorption process then proceeds due to the electrostatic attraction between these two counterions [18].

$$\underset{\text{chitosan}}{-NH_2} \underset{\text{Protonation}}{\overset{H^+}{\rightleftharpoons}} -NH_3^+$$

$$\text{dye-}SO_3Na \underset{\text{dissociation}}{\overset{H_2O}{\rightleftharpoons}} \text{dye}-SO_3^- + Na^+$$

$$\underset{\text{chitosan}}{-NH_3^+} + \text{dye-}SO_3^- \underset{\text{electrostatic interaction}}{\rightleftharpoons} -NH_3^+ \; {}^-O_3S - \text{dye}$$

Fig. 1 Mechanism of anionic dye adsorption by chitosan under acidic conditions

2.2 *Azo Dye Wastewater Treatment Using Combined Physical–Biological Processes*

Treatment of azo dye-containing wastewater using combined physical–biological processes has been the subject of a large number of research papers. Sirianuntapiboon and Sansak [19] developed a combined granular activated carbon (GAC) and sequencing batch reactor (SBR) system to treat both synthetic and raw textile wastewater containing direct dyes (direct blue 201 and direct red 23) under various concentrations of biosludge and dyestuffs. Ong et al. [20] investigated the feasibility of decolorization of azo dye Acid Orange 7-containing wastewater using a GAC-biofilm configured packed column system.

A combination of membrane technologies and biological treatment processes has been extensively applied in azo dye wastewater treatment. In the previous works, different membrane processes were used in the treatment of a biologically treated textile wastewater, such as microfiltration, ultrafiltration, and nanofiltration (NF) [21, 22] and the results showed that the reuse of the treated water was possible. You et al. [23] developed a combination of anaerobic–oxic membrane bioreactor and reverse osmosis (RO) processes to treat the synthetic textile dyeing wastewater (Reactive Black 5, RB5). The results indicated that the anaerobic tank can enhance the chemical oxygen demand (COD) and true color removal, while the RO unit can further remove the true color. The anaerobic biodegradation of azo dyes was successfully applied to decolorization of the concentrations from the NF treatment of real textile effluents [24, 25]. The anaerobic phase was followed by aerobic oxidation, which aimed at the destruction of the aromatic amine released from azo dye, and the aromatic amine was completely degraded in the aerobic reactor [26]. In another work, Lu et al. [27, 28] developed a combined process of biological process and sub-filter technology to treat printing and dyeing wastewater. The results showed that final effluent quality satisfied the requirement of water quality for printing and dyeing process.

Based on the previous studies and the above discussions, it is required to achieve environmentally sustainable development in textile industry since closing of water cycle is highly recommended [29]. It can often be done by the application of the combined membrane–biological processes, especially by the implementation of NF–biological and RO–biological processes, which enables the water reuse process, thereby the minimization of freshwater consumption.

3 Combined Chemical–Biological Processes

Chemical processes include reduction and oxidation. Conventional chemical (coagulation–flocculation) and advanced oxidation processes (AOPs), such as chemical oxidation (ozonation, Fenton oxidation, Fe^{2+}/H_2O_2), ultrasonic chemical oxidation, photocatalysis oxidation (UV/H_2O_2, UV/O_3, and $UV/O_3/H_2O_2$), electrochemical oxidation, and irradiation oxidation have been used to treat refractory textile industry effluents [30, 31]. They often combined with conventional biological processes. In fact, many studies have recently reported the combination of chemical processes and anaerobic or anoxic biological process for treatment of textile industry effluents.

3.1 Oxidation Mechanism for Azo Dyes Removal

Among these physical–chemical processes, Fenton's oxidation is one of the oldest AOPs, which are used successfully as it is comparatively cheap and easy to handle reagents. In Fenton's system, the Fenton's regent reacts with H_2O_2 to generate •OH as shown in the Fig. 2 [32].

A mixture of hydrogen peroxide and ferrous iron is effective for color and COD removal of dye effluent, which is effective for complete color removal and partial degradation of organic matter.

Among the AOPs, heterogeneous photocatalysis appears as an interesting technique for the treatment of azo dyes. Indeed, titan dioxide (TiO_2) activation under UV irradiation ($l < 390$ nm) allows the generation of highly reactive free radicals •OH from water or hydroxide ions. These free radicals can then react with the persistent components adsorbed on the surface of TiO_2 until their total mineralization. The photocatalytic mechanisms of TiO_2 are assumed as follow (Fig. 3) [33].

The ambient temperature and the possible use of solar UV are the advantages of photocatalysis; moreover, TiO_2 is not toxic. The reaction mechanisms of TiO_2 photocatalytic oxidation of azo dyes was similar to the biodegradation process of oxidation of azo dyes with •OH radical.

One of the reaction mechanisms of oxidation of azo dyes that react with •OH radical was proposed as follows (Fig. 4) [34].

$$Fe^{2+} + H_2O_2 \rightarrow Fe^{3+} + HO^- + \cdot OH \quad (1)$$

$$Fe^{3+} + H_2O_2 \rightarrow Fe^{2+} + HO_2 \cdot + H^+ \quad (2)$$

$$2H_2O_2 \rightarrow \cdot OH + HO_2 \cdot + H_2O \quad (3)$$

Fig. 2 Fenton's oxidation mechanism for azo dyes removal

$$TiO_2 + hv \rightarrow e^- + h^+ \quad (1)$$

$$h^+ + HO^- \rightarrow \cdot OH \quad (2)$$

$$\cdot h^+ + HO_2 \rightarrow H^+ + \cdot OH \quad (3)$$

$$e^- + Ti^{4+} \rightarrow Ti^{3+} \quad (4)$$

$$TiO_2 + e^- + O_2 \rightarrow \cdot O_2^- + TiO_2 \quad (5)$$

$$2OH\cdot \rightarrow H_2O_2 + O_2\cdot \quad (6)$$

Fig. 3 Photocatalytic mechanisms of TiO_2 for azo dyes removal

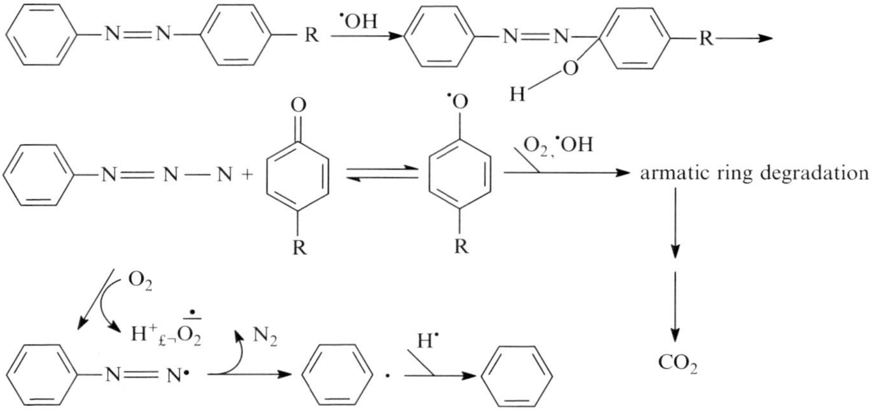

Fig. 4 The reaction mechanism of oxidation of azo dyes which react with ·OH radical

3.2 *Combined Chemical Coagulation–Biological Processes*

The effectiveness of a combined reduction–biological treatment system for the decolorization of nonbiodegradable textile dyeing wastewater has been investigated. The bench-scale experimental comparison of this technique with other reported combined chemical–biological methods showed higher efficiency and lower cost for the new technique [35].

3.3 *Combined Advanced Oxidation–Biological Processes*

In recent years, AOP followed by biological treatment is emerging as a potential process to pretreat azo dye-containing effluents, since the chromophore groups with conjugated double bonds, which are responsible for color, can be broken down by AOP either directly or indirectly forming smaller molecules, thereby improving the biodegradability of azo dye-containing wastewaters.

Ozonation is capable of decomposing the highly structured azo dye molecules into smaller ones, which can easily be biodegraded in an activated sludge process (ASP). The treatment efficiency of the aqueous Acid Red-151 (AR151) solutions using pre-ozonation followed by ASP has been investigated. Results showed that the pre-ozonation process can enhance the biodegradability of the azo dyes, and the treatment efficiency of biological treatment process could be higher [36]. Lu et al. [27, 28] work on the treatment of wastewater containing azo dye reactive brilliant red X-3B using sequential ozonation and up-flow biological aerated filter process. The experimental results showed that the combined process was a promising technique to treat wastewater containing azo dye.

Though Fenton's reagent is capable of dearomatization of dyestuff, there exists many problems such as the generation of aromatic amines, high reagent costs, and production of sludge which contain high amount of Fe (III), which need to be treated by safe disposal methods. Therefore, there is need for further research for finding an alternative economical treatment method for complete mineralization of textile azo dyes. Idil et al. [37] studied the effect of Fenton-treated acid dyes and a reactive dye on aerobic, anoxic, and anaerobic processes. The results indicated that Fenton process can be recommended for complete color and partial organic carbon removal. Marco et al. [38] developed Fenton's reagent/aerobic biological sequential processes to decolorize aqueous azo dye RB5. Color removal efficiency of about 91% for an initial RB5 concentration of 500 mg/L was achieved. Tantak et al. [39] found that Fenton's oxidation process followed by aerobic sequential batch reactors (SBRs) was viable method for achieving significant biodegradation of azo dyes. Biodegradation of a commercial homo-bireactive dye (Procion Red H-E7B, 250 mg/L) using combined photo-Fenton reaction with an aerobic SBR has been carried out [40]. The best results were obtained with 60 min of 10 mg/L Fe (II) and 125 mg/L H_2O_2 photo-Fenton pretreatment and 1 day hydraulic retention time (HRT) in SBR.

In heterogeneous photocatalytic process, stable organic compounds with high molecular weights are broken down into smaller structures, which are more biodegradable. However, a long period of time can be required during photocatalytic degradation. More recently, the integration of two processes, photocatalysis and biological treatment, was extensively applied in the treatment of azo dye-containing effluents [41–44].

Electrochemical methods, applied as a polishing treatment to textile effluents to eliminate persistent organic compounds, have been receiving great attention in the last years. Carvalho et al. [45] studied the electrochemical oxidation of the metabolites of Acid Orange 7 after anaerobic biotreatment, which obtained in experiments carried out in an up-flow anaerobic sludge blanket (UASB) reactor. Results have shown an almost complete elimination of the persistent pollutants and a COD removal higher than 70%. In another work, an electroenzymatic method that uses an immobilized horseradish peroxidase enzyme was investigated to degrade orange II (azo dye) within a two-compartment packed-bed flow reactor. It was found that removal of orange II was partly due to its adsorption to the graphite felt. The overall application of the electroenzymatic led to a greater degradation rate than with the use of electrolysis alone [46].

4 Combined Anaerobic–Aerobic Treatment Processes

4.1 Mechanism of Decolorization

Azo dye molecules have color due to their azo bond, auxochromes, and system of conjugated double bonds. The azo bond, while resistant to aerobic degradation, can be cleaved under anaerobic or anoxic condition, resulting in decolorization and the production of aromatic amines. Anaerobic reduction of the azo dyes is relatively easy to achieve, but the products have been found to be biorecalcitrant

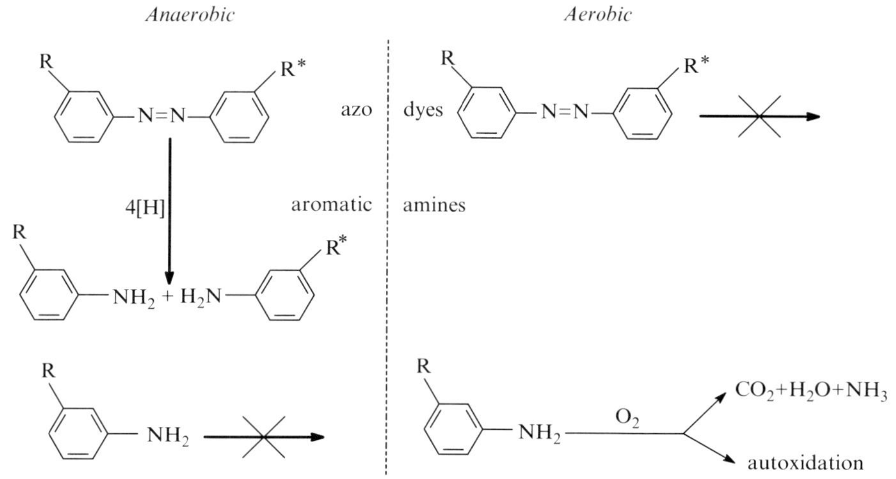

Fig. 5 Proposed pathway of azo dyes biodegradation in bioreactores

under anaerobic conditions [47]. Therefore, bacterial azo dye biodegradation proceeds in two stages. The first stage involves reductive cleavage of the dyes' azo linkages, resulting in the formation of – generally colorless but potentially hazardous – aromatic amines. The second stage involves degradation of the aromatic amines, while bacterial biodegradation of aromatic amines is an almost exclusively aerobic process (Fig. 5). As above the fact that azo dyes are decolorized under anaerobic or anoxic condition and many aromatic amines are completely degraded under aerobic conditions, the anaerobic–aerobic sequential process might be effective in achieving the complete treatment of azo dye-containing wastewaters.

4.2 Combined Anaerobic–Aerobic Treatment of Azo Dye-Containing Wastewaters in Bioreactors

As stated earlier, the biodegradation of azo dyes requires an anaerobic and aerobic phase for the complete mineralization. The required condition can be implemented either by spatial separation of the two sludge using a sequential anaerobic–aerobic reactor system or in one reactor in the so-called integrated anaerobic–aerobic reactor system. In recent years, combined anaerobic–aerobic treatment technologies are extensively applied in the treatment of azo dye-containing wastewaters. Table 1 lists the systems based on combined anaerobic–aerobic treatment in separate reactors. Table 2 lists SBR based on temporal separation of the anaerobic and the aerobic phase. Table 3 lists the other systems, either hybrids with aerated zones or micro-aerobic systems based on the principle of limited oxygen diffuse in microbial biofilms [91].

Table 1 Treatment of azo dye-containing wastewater using sequential anaerobic–aerobic reactor systems

Anaerobic		Aerobic		Wastewater characteristics				Color removal		Aromatic amines			References
Type[a]	HRT (h)	Type[b]	HRT (h)	ww[c]	Dye[d]	Conc. (mg/L)	Substrates[e]	Anaerobic (%)	Aerobic[f]	Recovery anaerboic[g]	Removal aerobic[h]	Detect method[i]	
3[j]	36	1[j]	36	S[k]	AO10, ma	5–100	Glucose	90–100	+	25–50%	Max. 100%	2	[48]
					ABk1, da	10–100		100					
					DR2, da	25–200		95–100					
					DR28, da	25–200		80–100					
1	24	1	19	S	h-RR141, da, mct	450	Starch and acetate	64	11%	+	+	1	[49]
1	24	1	19	S	h-RR141, da, mct	150–750	Starch and acetate	38–59	6.82%	n.e.	n.e.		[50]
1	24–28	1	NM	S	h-RR141, da, mct	1,500	Starch and acetate	~78	−7%	n.e.	n.e.		[51]
5	34–84	1	NM	S	h-RR141, da, mct	1,500	Starch and acetate	Max. 62	−7%	n.e.	n.e.		[51]
2	24	4	NM	S	AO7, ma	5–40	ME, peptone, YE, chow	20–90[l]	0	+	n.e.	n.m.	[52]
					AO8, ma	5–40			0				
					AO10, ma	5–40			0				
					AR14, ma	5–40			+				
2	31	4	3.1	S	AO10, ma	10	ME, peptone, YE, chow	~62	+	<1%	+	1-MS	[53]
					AR14, ma	10		~90					
					AR18, ma	10		~90					
4	15	2	7.5	S	h-RV5, ma, vs	650–1,300	Acetate and YE	90–95	–	69–83%	~100%[m]	1	[54]
4	31	2	7.5	S	h-RBk5, ma, vs	600	Acetate and YE	~70	+	n.e.	–[n]	4	[55]
4	15	2	7.5	S	(h-)RBk5, ma, vs	530	Acetate and YE	~100	−35%[o]	100%[p]	+[p]	1-MS	[56]
1	820	1[q]	23	S	AY17, ma	40	Glucose	20	0–13%	n.e.	n.e.		[57]
					(BB3, ox)	40		–72					
					BR2, az	40		–78					

1	6–10	1	6.5	R/S	Mixed with simulated municipal wastewater			70–80	10–20%	$+^{r}$	$+^{r}$		[57]
4	7–8	2	4.5–5	R	Textile dye wastewater with PVA and LAS as main COD			60–85		n.e.	n.e.		[58]
1	6–10	1	6	R	Textile dye wastewater with PVA and LAS as main COD			90–95	Max. 96%	n.e.	n.e.		[59]
3	6	3(2)	7.7–8.6	S	AY17, ma	25	Starch and acetate	0	+	n.e.	n.e.		[60]
					BR22, ma	200		>99					
1	25	1	10	S	MY10, ma	100–200	Ethanol	~100	0	~100%^{s}	~100%^{s}	1	[61]
3^{t}	Var.	3^{t}	12–24	S	DisB79, ma	25–150	Glucose, acetate none	Max. 100	n.m.	~40%^{u}	65%^{v}	1, 2, 3	[62]
3	^{w}	3	^{w}	S	h-RR198, ma, vs + mct	5,000	Starch	97	1.97	2–3%	~100%	1	[63]
1^{x}	24–48	1^{p}		R	Highly colored textile wastewater			70–90	−100%	n.e.	n.e.		[64]
1	15–16.5	1	55–60	S	DBk38, ta	100–320	Glucose	80–100	+	85–95%	~50%	2	[65]
1	86.4	1	432	S	DBk38, ta	100	Glucose	81	13%	74%	81%	2^{y}	[66]
1	3–30	1	10–30	S	RBk5, da, vs	100	Glucose	82–98	n.m.	n.e.	n.e.		[67]
1	3–30	1	10–108	S	RBk5, da, vs	100	Glucose	82–98	–	n.e.	n.e.	3	[68]
1	2.6–26	1	10–102	S	DR28,da	100–400	Glucose	97–100	n.m.	40–95%	80–100%	2	[69]
1	3–30	1	10–108	S	RBk5, da, vs	100	Glucose	87–98	−10 to 20%	n.e.	n.e.		[70]
1	2.5–19	1	9–67	S	DR28, da	100	Glucose	92–97	−1 to 15%				
1	10	1	108	R	Cotton mill wastewater (CMW)			46–55	−10 to 25%	+	35–90%	1, 2	[71]
					CMW + mixture of azo dye (250–500 mg/L) and glucose			60–75	−1 to 15%	+	40–80%		
3	12–72	1	10	S	R195, ma, vs + mct	50–400	Molasses	60–100^{c}	Max. 15%	n.e.	n.e.		[72]
3^{z}	12–72	1	10	R	Textile wastewater with added glucose and nutrients			60–85	10%	n.e.	n.e.		[73]
1	24	5	21.5	S	AO7, ma	60–300	Glucose + peptone	60–97	+	n.e.	n.e.		[74]
(6)	15–18	1	15–18	R	Bleaching, scouring, (desizing) wastewater containing10–15 g/L dyes			50–70	−5 to 5%	n.e.	n.e.		[75]
1	7							80–95					

(*continued*)

Table 1 (continued)

Anaerobic		Aerobic		Wastewater characteristics				Color removal		Aromatic amines			References
Type[a]	HRT (h)	Type[b]	HRT (h)	ww[c]	Dye[d]	Conc. (mg/L)	Substrates[e]	Anaerobic (%)	Aerobic[f]	Recovery anaerboic[g]	Removal aerobic[h]	Detect method[i]	
3	26–90	–6	480	R	1. Reactive dyebath waste and ww with starch and PVA			89–94	1–2%	n.e.	n.e.		[76]
					2. Split flows from yarn processing			81–92	1–7%				

[a]Anaerobic reactor types: 1, up-flow anaerobic sludge bed; 2, anaerobic fluidized bed; 3, anaerobic filter; 4, anaerobic rotating disc; 5, inclined tubular digester; (6, pre-acidification tank)

[b]Aerobic reactor: 1, aerobic tank; 2, aerobic rotating disc; 3, aerobic filter; 4, swisher; 5, sequential batch reactor; 6, aerobic biodegradability tests (BOD_{20})

[c]Wastewater type (ww): S, synthetic wastewater; R, real wastewater

[d]Dyes: First abbreviation refers to Colour Index Generic Names: A, acid; B, basic; D, direct; Dis, disperse; M, modant; R, reactive; B, blue; Bk, black; O, orange; R, red; V, violet; Y, yellow. Second abbreviation refers to amount of azo linkages: ma, monoazo; da, disazo; ta, triazo; (ox, oxazine; az, azine). Third abbreviation refers to reactive groups (reactive dyes only): vs, vinylsulfone; mct, monochlorotriazine. The prefix "h-" means hydrolyzed (reactive dyes only). Libra et al. [56] investigated both hydrolyzde, partially hydrolyzed, and nonhydrolyzed Reactive Black 5

[e]Substrates: YE, yeast extract; PVA, polyvinylalcohol; LAS, linear alkyl benzene sulfonate; ME, meat extract

[f]Color removal aerobic: positive values express the additional color removal as percentage of the influent color, negative values express development of color (autoxodation) as percentage of influent color. "n.m." not mentioned

[g]Anaerobic aromatic amine recovery: "+" indicates nonquantified sign of recovery; "n.e." not evaluated

[h]Aerobic aromatic removal; "+" indicates nonquantified sign of removal; percentages express removal of recovered aromatic amines; "n.e." not evaluated

[i](Main) detection method aromatic amines:1, HPLC; 1-MS; 2, diazotization-based colorimetric method; 3, UV spectrophotometry ; 4, DOC measurements

[j]Both anaerobic and aerobic reactor inoculated with a mixture of four pseudomonads isolated from dyeing effluent-contaminated soils

[k]Nitrogen-free medium

[l]Depending on dye concentration and HRT. All dyes >80% decolorization at high HRT

[m]Complete removal of the metabolites from anaerobic treatment, probably mostly due to autoxidation

[n]Presumably no removal of dye metabolites: hardly any DOC removal and only slight decrease of toxicity

[o]Data refer to fully hydrolyzed RBk5, less color removal for partially hydrolyzed RBk5

[p]Fully hydrolyzed RBk5 was completely converted in the anaerobic phase, to *p*-aminobenzene-2-hydroxyleethylsulfonic acid (2 mol *p*-ABHES per mol RBk5) and 1,2,7-triamino-8-hydroxynaphthalene-3-6-disulfonic acid (1 mol TAHNDS per mol RBk5). In the aerobic phase, *p*-ABHES was mineralized while TAHNDS autoxidized to 1, 2-ketimino-7-amino-8-hydroxynaphthalene-3-6-disulfonic acid. Partially hydrolyzed RBk5 was not completely converted in the anaerobic phase. *p*-ABHES and TAHNDS were detected, but in relatively small amounts. There was no removal of *p*-ABHES in the aerobic phase

[q]Semi-continuous system

[r]Increased BOD_5/COD ratio after anaerobic treatment may point at formation of biodegradable dye metabolites

[s]Almost complete recovery of the dye metabolites, sulfanilic acid; partial anaerobic degradation of the other, 5-aminosaliculate. In the aerobic reactor complete mineralization of 5-aminosalicylate;after bioaugmentation also complete mineralization of sulfanilic acid
[t]Discontinuously fed reactors
[u]Percentage expresses HPLC recovery of 2-bromo-4,6-dinitroaniline (BDNA). Additional thin layer chromatography measurements indicate anaerobic transformation BDNA
[v]Percentage based on total amine measurements (diazotization method)
[w]HRT total system 96 h
[x]Sludge bed amended with granular activated carbon
[y]Additional support of aerobic AA removal from HPLC-MS and nitrate analyses
[z]Inoculated with a facultative anaerobic consortium (mixture of *Alcaligenes faecolis* sp. and *Comamonas acidourans* sp.)
1-MS = HPLC-MS

Table 2 Treatment of azo dye-containing wastewater using anaerobic–aerobic sequenced batch reactor (SBR) systems

Cycle			Wastewater characteristics				Color removal		Aromatic amines			References
Anaerobic (h)	Aerobic (h)	Total time (h)	ww[a]	Dye[b]	Conc. (mg/L)	Substrates[c]	Anaerobic	Aerobic[d]	Recovery anaerobic[e]	Removal aerobic[f]	Detect method[g]	
13	8	24	S	h-RV5, ma, vs	60–100	Starch	30–90%[h]	+/0	+	+[i]	1	[77]
9–12	8–12	24	S	h-RV5, ma, vs	60–100	Starch	20–90%[j]	n.m.	+	+[i]	1	[78]
				h-RBK5, da, vs	30		65%[k]		n.e.	n.e.		
9–13	8–12	24	S	h-RV5, ma, vs	100	Starch	Max. 90%	n.m.	+	+[i]	1	[79]
10.5	10	24	S	h-RV5, ma, vs	100	Starch	90–99%	n.m.	+	n.e.	1	[80]
10.5–17	3.5–10	24	S	AO7, ma	25	Starch	5–55%		n.e.			
10.5	10	24	S	AO7, ma	25	Starch + lactate	Max.95%[l]	+	n.e.	+	3	[81]
0–12	8–12	24	R	Wool dyeing effluent with azo and anthraquinone dyes			+	+	n.e.	−	1	[82]
18	5	24	S	RBK5, ma, vs	20–100	Glucose and acetate	58–63%		+			
				(RB19, aq, vs)	20–100		(64–32%)					
				(RB5, aq, mct)	20–100		(66–41%)					
				(RB198, ox, hh)	20–100		(−)[m]					
18	5	24	S	RBK5, da, vs	10	NB + acetate or glucose	68–72%	2–8%	+	n.e.	3	[83]
	3–1						63–68%	8–11%				
0–8	0.5	12	S	RBK5, da, vs	10–80	NB + acteate or NB + glucose	30–61%	2–17%	+	n.e.	3	[84]
18.5		24	S	h-RBK5, da, vs	533	Starch, PVA, CMC	86–96%	+	+	+/0	3	[85]

[a]Wastewater type (ww): S, synthetic wastewater; R, real wastewater
[b]Dyes: first abbreviation refers to Colour Index Generic Names: A, acid; R, reactive; B, blue; BK, black; O, orange; V, violet. Second abbreviation refers to amount of azo linkages: ma, monoazo; da, disazo; (aq, anthraquinone; ox, oxazine). Third abbreviation refers to reactive groups (reactive dyes only): vs, vinylsulfone; mct, monochlorotriazine; hh, halogenohetrocyclic. The prefix "h-" means hydrolyzed (reactive dyes only)
[c]Substrates: NB, nutrient broth; PVA, polyvinyl alcohol; CMC, carboxymethylcellulose
[d]Color removal aerobic: positive values express the additional color removal as percentage of the influent color, negative values express development of color (autoxidation)as percentage of influent color. "n.m." not mentioned
[e]Anaerobic aromatic amine recovery: "+" indicates nonquantified sign of recovery; "n.e." not evaluated
[f]Aerobic aromatic amine removal: "+" indicates nonquantified sign of removal; "p" nonquantified sign of partial removal; percentages express removal of recovered aromatic amines; "n.e." not evaluated
[g](Main)detection method aromatic amines: 1, HPLC; 3, UV spectrophotometry
[h]~90% color removal at a sludge concentration of 2.0 g VSS/L and SRT = 15 days, ~30% color removal at a sludge concentration of 1.2 g VSS/L and SRT = 10 days
[i]No degradation of RV5's constituent naphthalene-based amine; (bio)transformation but no mineralization of its benzene-based amine
[j]~90% color removal at a sludge concentration = 2.0 g VSS/L, SRT = 15 days and feed dye concentration = 60 mg/L, ~20% color removal at a sludge concentration = 1.2 g VSS/L, SRT = 10 days and feed dye concentration = 100 mg/L
[k]No effect of changing the SRT
[l]Highest color removal achieved with addition of anthraquinone-2,6-disulfonate
[m]Could not be quantified

Table 3 Treatment of azo dye-containing wastewater using integrated anaerobic–aerobic reactor systems

System	Wastewater characteristics					Color removal		Aromatic amines			References
Reactor type[a]	Total time (h)	ww[b]	Dye[c]	Conc. (mg/L)	Substrates[d]	Anaerobic (%)	Aerobic	Recovery anaerobic[e]	Removal aerobic[f]	Detect method[g]	
EGSB with oxygenation of recycled effluent	36–43	S	MY10, ma	59–65	Ethanol	~100		+[h]	+[h]	1	[86]
	26–34		4-PAP, ma	50		<100[i]		+[h]	+[h]	1	
UASB with aerated upper part	1–100	S	DY26, da	300	Ehanol	40–70	9.8	+[j]	+[j]	3	[87]
RAD	0.16–3		AO7, ma			18–97[k]				1, 5	[88]
RAD	2	S	AO8, ma	0–22	ME, peptone, YE, trout chow	20–90[k]		+	+		[89]
		S	AO10, ma	n.m.	ME, peptone, YE, trout chow	Max. 60		n.e.	n.e.		
			AR14, ma		Starch, PVA, CMC	Max. 60					
Baffled reactor with anaerobic and aerobic compartments	48 + 18	S	h-RBk5, da, vs	500		84–88		+	+[l]	3	[90]

[a]Reactor types: EGSB, expanded granular sludge bed; UASB, up-flow anaerobic sludge blanket; RAD, rotating annular drum

[b]Wastewater type (ww): S, synthetic wastewater

[c]Dyes: 4-PAP is 4-phenylazophenol. For the other dyes, the first abbreviation refers to Colour Index Generic Names: A, acid; D, direct; M, mordant; R, reactive; Bk, black; O, orange; R, red; Y, yellow. The second abbreviation refers to the amount of azo linkages: ma, monoazo; da, disazo. The third abbreviation refers to the reactive groups (reactive dyes only): vs, vinylsulfone. The perfix "h-" means hydrolyzed (reactive dyes only)

[d]Substrates: YE, yeast extract; PVA, polyvinyl alcohol; ME, meat extract; CMC, carboxymethylcellulose

[e]Anaerobic aromatic amine recovery: "+" indicates nonquantified sign of recovery; "n.e." not evaluated

[f]Aerobic aromatic amine removal: "+" indicates nonquantified sign of removal; percentages express removal of recovered aromatic amines; "n.e." not evaluated

[g](Main)detection method aromatic amines: 1, HPLC; 3, UV spectrophotometry; 5, GC-MS

[h]Aromatic amines from MY10: almost complete recovery of sulfanilic acid, partial anaerobic degradation of 5-aminosalicylate; aromatic amines from 4-PAP: complete mineralization of aniline, autoxidation of 4-aminophenol

[i]Residual color due to autoxidation of 4-aminophenol (one of 4-PAP's constituent aromatic amines)

[j]One of the dye's aromatic amine (5-aminosalycilate) was partially degraded in the anaerobic part and underwent autoxidation in the aerobic part

[k]At high oxygen/low COD flux, dye removal probably (partly) due to aerobic degradation

[l]Decrease of toxicity after addition of adapted biomass may indicate biological degradation of aromatic amines

4.3 Effect of Bioreactors on Azo Dyes Biodegradation

Because of the highly variable nature of biological treatment systems and especially textile effluents, there are many factors that may affect the biodegradation rate of azo dyes. Throughout the literature, researchers have discussed various problems associated with azo dyes biodegradation that may or may not be anticipated or remedied. Non-dye related parameters such as temperature, pH, and HRT, dissolved oxygen (DO) or nitrate concentration, type and source of reduction equivalents, bacteria consortium, and cell permeability can all affect the biodegradation of azo dyes. Dye related parameters such as class and type of azo dye (i.e., reactive-monoazo), reduction metabolites, dye concentration, dye side-groups, and organic dye additives could also affect the biodegradability of azo dye-containing wastewaters.

The azo dye structure plays a significant role in the azo dye biodegradation rate. Depending on the number and placement of the azo linkages, some dyes will biodegrade more rapidly than others. Brown and Laboureur [92] found that poly-azo dyes were less likely to degrade than mono- or diazo dye types. Suzuki et al. [93] provided a correlation of aerobic biodegradability of 25 sulfonated azo dyes with their chemical structures. In another work, the biodegradation of azo dyes by algae was studied and found that the reduction rate of azo dyes was related to the molecular structure of the dye and species of algae used [94]. In general, the more azo linkages that must be broken will cause the reduction rate to be slower.

Several studies have reported a positive relationship between the hydraulic retention time of the anaerobic stage and the color removal efficiency [95].

Another important factor to evaluate is the initial dye concentration of the azo dye-containing wastewaters. Swshadri and Bishop drew a conclusion that dye concentration may cause a drop in the percentage of dye removal. Furthermore, the inhibition may be directly related to the effects of increased dye metabolite formation due to higher dye concentrations. Cariell et al. [96] found that C.I. Reactive Red 141 was inhibitory to anaerobic organisms at concentrations greater than 100 mg/L.

The wastewater pH value can affect the proper function of both anaerobic and aerobic organisms [97]. Wuhrmann et al. [98] investigated the effect of pH on dye reduction rates. They stated that an exponential increase in the decolorization rate was observed by decreasing the pH. Furthermore, wastewaters from textile processing and dyestuff manufacture industries contain substantial amounts of salts in addition to azo dye residues. Muhammas and Crowley [99] found an inverse relationship between the velocity of the decolorization reaction and salt concentration. Therefore, biological treatment system generally require pretreatment of the azo dye-containing wastewaters to dilute high salt concentrations or screen salt-tolerant bacteria.

Nitrate and oxygen also may play an important role in determining the rate of azo dyes reduction. Wuhrmann et al. demonstrated that obligate aerobes might actually decolorize azo dye compounds under temporary anoxic conditions. However, high nitrite concentrations in the mixed liquor of activated sludge plants could significantly inhibit dye removal.

5 Conclusion

The state of the art in the field of integrated processes for treatment of azo dye-containing wastewater was reviewed in this paper, based on a substantial number of relevant references published recently, and the following conclusions were reached:

1. The above work indicated that the combined processes for the treatment of azo dye-containing wastewater have become promising alternatives to replace conventional technologies used for the purposes of decolorization. These processes are efficient in azo dyes removal with advantages of being cheap, nontoxic, and biocompatible.
2. There are abundant literatures concerning the treatment of azo dye-containing wastewaters using combined advanced oxidation–biological processes. Among them, an integrated technique using photocatalysis and sequential biological treatment was extensively applied.
3. The reductive cleavage of azo compounds to aromatic amines requires anaerobic conditions and then bacterial biodegradation of the aromatic amines is an almost exclusively aerobic process; therefore, a combined anaerobic–aerobic bacterial process is most effective for removing azo dyes from wastewater.

Although extensive work has been done, future research needs to look into some of the following aspects:

Reaction mechanism of azo dyes removal: Limited study has focused on the reaction mechanism of azo dyes removal. The research of mechanistic and mathematical models to optimize the integrated process and to characterize the interaction between the reactant and azo dyes should be carried out in the future.

Real effluent: The real wastewater containing azo dyes are proposed to treat using integrated processes on the basis of thermodynamics and reaction kinetics studies.

Large-scale experiments: Integrated processes are basically at the stage of laboratory-scale study in spite of unquestionable progress. Much work in this area is necessary to demonstrate the possibilities on an industrial scale.

Acknowledgments The authors gratefully acknowledge Dr. F.P. van der Zee and S. Villaverde (Combined anaerobic–aerobic treatment of azo dyes – a short review of bioreactor studies), whose work was much helpful for us.

References

1. Stolz A (2001) Basic and applied aspects in the microbial degradation of azo dyes. Appl Microbiol Biotechnol 56:69–80
2. Vijaykumar MH, Vaishampayan PA, Warren SH et al (2007) Decolourization of naphthalene-containing sulfonated azo dyes by Kerstersia sp strain VKY1. Enzyme Microb Technol 40:204–211

3. Aleboyeh A, Olya MA, Aleboyeh H (2009) Oxidative treatment of azo dyes in aqueous solution by potassium permanganate. J Hazard Mater 162:1530–1535
4. Lodha B, Chuaudhari S (2007) Optimization of Fenton-biological treatment scheme for the treatment of aqueous dye solutions. J Hazard Mater 148:459–466
5. Zhang H, Duan L, Zhang Y et al (2005) The use of ultrasound to enhance the decolorization of the C.I. Acid Orange 7 by zero-valentiron. Dyes Pigm 65:39–43
6. Kulla HG, Klausener F, Meyer U et al (1983) Interference of aromatic sulfo groups in the microbial degradation of the azo dyes Orange I and Orange II. Arch Microbiol 135:1–7
7. Matthew EJ, Spalding JW, Tennant RW (1993) Transformation of BALB/c-3T3 cells via transformation responses of 168 chemicals compared with mutagenicity in salmonella and carcinogenicity in rodent bioassays. Environ Health Perspect 101:347–482
8. Chung KT, Stevens SEJ (1993) Degradation of azo dyes by environmental microorganisms and helminthes. Environ Toxicol Chem 12:2121–2132
9. Weisburger JH (2002) Comments on the history and important of aromatic and heterocyclic amines in public health. Mutat Res 506–507:9–20
10. Hao OJ, Kim H, Chiang PC et al (2000) Decolorization of wastewater. Crit Rev Environ Sci Technol 30:449–502
11. Maier J, Kandelbauer A, Erlacher A et al (2004) A new alkali-themostable azo reductase from Bacillus sp. strain SF. Appl Environ Microbiol 70:837–844
12. Scott JP, Ollis DF (1995) Integration of chemical and biological oxidation processes for water treatment: review and recommendations. Environ Prog 14:88–103
13. Scott JP, Ollis DF (1997) Integration of chemical and biological oxidation processes for water treatment: II. Recent illustrations and experiences. J Adv Oxid Technol 2:374–381
14. Carriere J, Mourato D, Jones D (1993) Answers to textile wastewater problems: membrane bioreactor systems. In: Proceedings of the international conference and exhibition, AATCC Book of Papers, Montreal
15. Cooper P (1993) Removing color from dyehouse wastewaters – a critical review of technology available. J Soc Dyers Color 109:97–100
16. Majewska-Nowak K (1992) Color removal by reverse osmosis. J Membr Sci 68:307–315
17. Mckay G (1980) Color removal by adsorption. Am Dyestuff Rep 69:38–45
18. Crini G, Badot PM (2008) Application of chitosan, a natural aminopolysaccharide, for dye removal from aqueous solutions by adsorption processes using batch studies: a review of recent literature. Prog Polym Sci 33:399–447
19. Sirianuntapiboon S, Sansak J (2008) Treatability studies with granular activated carbon (GAC) and sequencing batch reactor (SBR) system for textile wastewater containing direct dyes. J Hazard Mater 159:404–411
20. Ong SA, Toorisaka E, Hirata M, Hano T (2008) Combination of adsorption and biodegradation processes for textile effluent treatment using a granular activated carbon-biofilm configured packed column system. J Environ Sci 20:952–956
21. Bes-Piá A, Mendoza-Roca JA, Roig-Alcover L et al (2003) Comparison between nanofiltration and ozonation of biologically treated textile wastewater for its reuse in the industry. Desalination 157:81–86
22. Fersi C, Gzara L, Dhahbi M (2005) Treatment of textile effluents by membrane technologies. Desalination 185:1825–1835
23. You SJ, Tseng DH, Deng JY (2008) Using combined membrane processes for textile dyeing wastewater reclamation. Desalination 234:426–432
24. Gomes AC, Goncalves IC, de Pinho MN, Porter JJ (2007) Integrated nanofiltration and upflow anaerobic sludge blanket treatment of textile wastewater for in-plant reuse. Water Environ Res 79:498–506
25. Żyłłaa R, Sójka-Ledakowicza J, Stelmachb E et al (2006) Coupling of membrane filtration with biological methods for textile wastewater treatment. Desalination 198:316–325
26. Katarzyna P, Anna KS, Stanisław L et al (2009) Integration of nanofiltration and biological degradation of textile wastewater containing azo dye. Chemosphere 75:250–255

27. Lu XJ, Liu L, Yang B, Chen JH (2009) Reuse of printing and dyeing wastewater in processes assessed by pilot-scale test using combined biological process and sub-filter technology. J Clean Prod 17:111–114
28. Lu XJ, Yang B, Chen JH et al (2009) Treatment of wastewater containing azo dye reactive brilliant red X-3B using sequential ozonation and upflow biological aerated filter process. J Hazard Mater 161:241–245
29. El Defrawy NMH, Shaalan HF (2007) Integrated membrane solutions for green textile industries. Desalination 204:241–254
30. Ledakowicz S, Solecka M, Zylla R (2001) Biodegradation, decolourisation and detoxification of textile wastewater enhanced by advanced oxidation processes. J Biotechnol 89:175–184
31. Rodriguez M, Sarria V, Esplugas S et al (2002) Photo-Fenton treatment of a biorecalcitrant wastewater generated in textile activities: biodegradability of the photo-treated solution. J Photochem Photobiol A Chem 151:129–135
32. Goi A, Trapido M (2002) Hydrogen peroxide photolysis, Fenton reagent and photo-Fenton for the degradation of nitrophenols: a comparative study. Chemosphere 46:913–922
33. Mills G, Hoffmann MR (1993) Photocatalytic degradation of pentachlorophenol on TiO_2 particles: identification of intermediates and mechanism of reaction. Environ Sci Technol 27:1681–1689
34. Spadaro JT, Isabelle L, Renganathan V (1994) Hydroxyl radical mediated degradation of azo dyes: evidence for benzene generation. Environ Sci Technol 28(7):1389–1393
35. Ghoreishi SM, Haghighi R (2003) Chemical catalytic reaction and biological oxidation for treatment of non-biodegradable textile effluent. Chem Eng J 95:163–169
36. Gökçena F, Özbelge TA (2006) Pre-ozonation of aqueous azo dye (Acid Red-151) followed by activated sludge process. Chem Eng J 123:109–115
37. Idil AA, Betul HG, Jens-Ejbye S (2008) Advanced oxidation of acid and reactive dyes: effect of Fenton treatment on aerobic, anoxic and anaerobic processes. Dyes Pigm 78:117–130
38. Marco SL, Albino AD, Ana S et al (2007) Degradation of a textile reactive Azo dye by a combined chemical-biological process: Fenton's reagent-yeast. Water Res 41:1103–1109
39. Tantak NP, Chaudhari S (2006) Degradation of azo dyes by sequential Fenton's oxidation and aerobic biological treatment. J Hazard Mater 136:698–705
40. García-Montaño J, Torrades F, García-Hortal JA et al (2006) Degradation of Procion Red H-E7B reactive dye by coupling a photo-Fenton system with a sequencing batch reactor. J Hazard Mater B134:220–229
41. Brosillon S, Djelal H, Merienne N et al (2008) Innovative integrated process for the treatment of azo dyes: coupling of photocatalysis and biological treatment. Desalination 222:331–339
42. Harrelkas F, Paulo A, Alves MM et al (2008) Photocatalytic and combined anaerobic-photocatalytic treatment of textile dyes. Chemosphere 72:1816–1822
43. Moziaa S, Morawskia AW, Toyodab M et al (2009) Application of anatase-phase TiO2 for decomposition of azo dye in a photocatalytic membrane reactor. Desalination 24:97–105
44. Sudarjanto G, Keller-Lehmann B, Keller J (2006) Optimization of integrated chemical-biological degradation of a reactive azo dye using response surface methodology. J Hazard Mater 138:B160–B168
45. Carvalho C, Fernandes A, Lopes A et al (2007) Electrochemical degradation applied to the metabolites of Acid Orange 7 anaerobic biotreatment. Chemosphere 67:1316–1324
46. Kim GY, Lee KB, Cho SH et al (2005) Electroenzymatic degradation of azo dye using an immobilized peroxidase enzyme. J Hazard Mater B126:183–188
47. Xu MY, Guo J, Sun GP (2007) Biodegradation of textile azo dye by Shewanella decoloration S12 under microaerophilic conditions. Appl Microbiol Biotechnol 76:719–726
48. Rajaguru P, Kalaiselve K, Palanivel M et al (2000) Biodegradation of azo dyes in a sequential anaerobic-aerobic system. Appl Microbiol Biotechnol 54:268–273
49. O'Neill C, Lopez A, Esteves S et al (2000) Azo-dye degradation in an anaerobic-aerobic treatment operating on simulated textile effluents. Appl Microbiol Biotechnol 53:249–254

50. O'Neill C, Hawkes FR, Kawkes DW et al (2000) Anaerobic-aerobic biotreatment of simulated textile effluent containing varied ratios of starch and azo dye. Water Res 34:2355–2361
51. O'Neill C, Hawkes FR, Kawkes DL et al (1999) Colour in textile effluents-source, measurement, discharge consents and simulation: a review. J Chem Technol Biotechnol 74:1009–1018
52. Seshadri S, Bishop PL, Agha AM (1994) Anaerobic/aerobic treatment of selected azo dyes in wastewater. Waste Manange 14:127–137
53. FitzGerald SW, Bishop PL (1995) Two stage anaerobic-aerobic treatment of sulfonated azo dyes. J Environ Sci Health A Tox Hazard Subst Environ Eng 30:1251–1276
54. Sosath F, Libra JA (1997) Purification of wastewaters containing azo dyes. Acta Hydrochim Hydrobiol 25:259–264
55. Wiesmann U, Sosath F, Borchert M et al (2002) Attempts to the decolorization and mineralization of the azo dye C.I. Reactive Black 5. Wasser Abwasser 143:329–336
56. Libra JA, Borchert M, Vigelahn L et al (2004) Two stage biological treatment of a diazo reactive textile dye and the fate of the dye metabolites. Chemosphere 56:167–180
57. An H, Qian Y, Gu XS et al (1996) Biological treatment of dye wastewaters using an anaerobic-oxic system. Chemosphere 33:2533–2542
58. Zaoyan Y, Ke S, Guangliiiang S et al (1992) Anaerobic-aerobic treatment of a dye wastewater by combination of RBC with activated sludge. Water Sci Technol 26:2093–2096
59. Jiangrong Z, Yanru Y, Huren A et al (1994) A study of dye wastewater treatment using anaerobic-aerobic process. Proceedings of the seventh international symposium on anaerobic digestion, Cape Town, South Africa, pp 360–263
60. Basibuyuk M, Forster CF (1997) The use of sequential anaerobic-aerobic processes for the biotreatment of a simulated dyeing wastewater. Environ Technol 18:843–848
61. Tan NGG, Field JA (2000) Environmental technologies to treat sulfur pollution principles and engineering. IWA Publishing, London, pp 377–392
62. Cruz A, Buitron G (2001) Biodegradation of Disperse Blue 79 using sequenced anaerobic-aerobic biofilters. Water Sci Technol 44:159–166
63. Sarsour J, Janitza J et al (2001) Biological degradation of dye-containing wastewater. Wasser Luft Boden (6):44–46
64. Kuai L, De Vreese I, Vandevivere P et al (1998) GAC-amended USAB reactor for the stable treatment of toxic textile wastewater. Environ Technol 19:1111–1117
65. Sponza DT, Isik M (2005) Reactor performances and fate of aromatic amines through decolorization of Direct Black 38 dye under anaerobic-aerobic sequential. Process Biochem 40:35–44
66. IsiK M, Sponza DT (2004) Monitoring of toxicity and intermediates of C.I. Direct Black 38 azo dye through decolorization in an anaerobic-aerobic sequential reactor system. J Hazard Mater 114:29–39
67. Sponza DT, Isik M (2002) Decolorization and azo dye degradation by anaerobic-aerobic sequential process. Enzyme Microb Technol 31:102–110
68. Sponza DT, Isik M (2002) Decolorization and inhibition kinetic of Direct Black 38 azo dye with granulated anaerobic-aerobic sequential process. Water Sci Technol 45:271–278
69. IsiK M, Sponza DT (2003) Aromatic amine degradation in a USAB/CSTR sequential system treating Conge red dye. J Environ Sci Health A Tox Hazard Subst Environ Eng 38:2301–2315
70. IsiK M, Sponza DT (2004) Decolorization of azo dyes under batch anaerobic and sequential anaerobic-aerobic conditions. J Environ Sci Health A Tox Hazard Subst Environ Eng 39:1107–1127
71. IsiK M, Sponza DT (2004) Anaerobic/aerobic sequential treatment of a cotton textile mill wastewater. J Chem Technol Biotechnol 79:1268–1274
72. Kapdan IK, Oztekin R (2003) Decolorization of textile dyestuff Reactive Orange 16 in fed-batch reactor under anaerobic condition. Enzyme Microb Technol 33:231–235

73. Kapdan IK, Alparslan S (2005) Application of anaerobic-aerobic sequential treatment system to real textile wastewater for color and COD removal. Enzyme Microb Technol 36:273–279
74. Ong SA, Toorisaka E, Hirata M et al (2005) Decolorization of azo dye in a sequential USAB-SBR system. Sep Purif Technol 42:297–302
75. Frijters CTMJ, Vos RH, Scheffer G et al (2004) Decolorizing and detoxifying textile wastewater in a full-scale sequential anaerobic-aerobic system. In: Guiot SR, vol 4. IWA Publishing, London, pp 2387–2390
76. Minke R, Rott U (2002) Investigation to anaerobic pre-treatment of highly colorly textile industry wastewater. Wasser Abwasser 143:320–328
77. Lourenco ND, Novais JM, Pinheiro HM (2000) Reactive textile dye colour removal in a sequencing batch reactive. Water Sci Technol 42:321–328
78. Lourenco ND, Novais JM, Pinheiro HM (2001) Effect of some operational parameters on textile dye biodegradation in a sequential batch reactor. J Biotechnol 89:163–174
79. Lourenco ND, Novais JM, Pinheiro HM (2003) Analysis of secondary metabolite fate during anaerobic-aerobic azo dye biodegradation in a sequential batch reactor. Environ Technol 24:679–686
80. Albuquerque MGE, Lopes AT, Serralheiro ML et al (2005) Biological sulphate reduction and redox mediator effects on azo dye decolourisation in anaerobic-aeobic sequencing batch reactors. Enzyme Microb Technol 36:790–799
81. Goncalves IC, Penha S, Matos M et al (2005) Evaluation of an integrated anaerobic-aerobic SBR system for the treatment of wool dyeing effluents. Biodegradation 16:81–89
82. Luangdilok W, Paswad T (2000) Effect of chemical structures of reactive dyes on color removal by an anaerobic-aerobic process. Water Sci Technol 42:377–382
83. Panswad T, Iamsamer K, Anotai J (2001) Decolorisation of azo-reactive dye by polyphosphate and glycogen-accumulating organisms in an anaerobic-aerobic sequencing batch reactor. Bioresour Technol 76:151–159
84. Panswad T, Iamsamer K, Anotai J (2001) Comparison of dye wastewater treatment by normal and anoxic+anaerobic-aerobic SBR activated sludge process. Water Sci Technol 43:355–362
85. Shaw CB, Carliell CM, Wheatley AD (2002) Anaerobic-aerobic treatment of coloured textile effluents using sequencing batch reactors. Water Res 36:1993–2001
86. Tan NCG (2001) Integrated and sequential anaerobic-aerobic biodegradation of azo dyes. PhD Thesis, Agri-technology and Food Sciences, Sub-department of environmental technology, Wageningen University, Wageningen, The Netherlands
87. Kalyuzhnyi S, Sklyar V (2000) Biomineralisation of azo dyes and their breakdown products in anaerobic-aerobic hybrid and USAB reactor. Water Sci Technol 41:23–30
88. Harmer C, Bioshop P (1992) Transformation of azo dye AO7 by wastewater biofilms. Water Sci Technol 26:627–636
89. Jiang H, Bishop DT (1994) Aerobic biodegradation of azo dyes in biofilms. Water Sci Technol 29:525–530
90. Gottlieb A, Shaw C, Smith A et al (2003) The toxicity of textile reactive azo dyes after hydrolysis and decolourisation. J Biotechnol 101:49–56
91. Van der Zee FP, Villaverde S (2005) Combined anaerobic-aerobic treatment of azo dyes – a short review of bioreactor studies. Water Res 39:1425–1440
92. Brown D, Laboureur P (1993) The degradation of dyestuffs: part I primary biodegradation under anaerobic conditions. Chemosphere 12:397–404
93. Suzuki T, Timofei S, Kurunczi L et al (2001) Correlation of aerobic biodegradability of sulfonated azo dyes with chemical structure. Chemosphere 45:1–9
94. Jing L, Hou T (1992) Degradation of azo dyes by algae. Microbiology 57:3144–3149
95. Rai HS, Singh PPS, Cheema TK et al (2007) Decolorization of triphenylmethane dye-bath effluent in an integrated two-stage anaerobic reactor. J Environ Manage 83:290–297
96. Cariell CM, Barclay SJ, Naidoo N et al (1995) Microbial decolourisation of a reactive azo dye under anaerobic conditions. Water SA 21:61–69

97. Grady CPL Jr, Daigger GT, Lim HC (1999) Biological wastewater treatment. Marcel Dekker, Inc., New York, NY, p 1076
98. Wuhrmann K, Ke R, Guangliang S et al (1980) Investigation on rate-determining factors in the microbial reduction of azo dyes. Eur J Appl Microbiol Biotechnol 9:325–338
99. Muhammas A, Crowley ADE (2008) Decolorization of azo dyes by Shewanella sp. under saline conditions. Appl Microbiol Biotechnol 79:1053–1059

Decolorization of Azo Dyes by White Rot Fungi

Emrah Ahmet Erkurt, Hatice Atacag Erkurt, and Ali Unyayar

Abstract White rot fungi (WRF) produce various isoforms of extracellular peroxidases (lignin peroxidase-LiP and manganese peroxidase-MnP) and phenoloxidases (laccases), which are involved in the degradation of lignin in their natural lignocellulosic substrates. This ligninolytic system of WRF is directly involved in the degradation of various xenobiotic compounds and dyes. Liquid fermentation or solid-state fermentation techniques can be used for enzyme production. Crude enzymes or purified enzymes of WRF can be used for decolorization of azo dyes. Repeated-batch decolorization technique is a new approach that can be used for decolorization. There are different procedures to determine the enzyme(s) responsible for decolorization. Single step isolation and identification procedure (SSIIP) is a new and simple method that can be used for detection of the enzyme responsible for biodegradation of azo dyes.

Keywords Azo dye, Biodegradation, Decolorization, Laccase, Peroxidase, White rot fungus

Contents

E.A. Erkurt (✉) and A. Unyayar
Department of Environmental Engineering, Faculty of Engineering, University of Mersin, Ciftliikkoy Campus, 33343 Mersin, Turkey
e-mail: eaerkurt@mersin.edu.tr

H. Atacag Erkurt
Department of Environmental Engineering, Faculty of Engineering, Cyprus International University, Haspolat, Nicosia, North Cyprus, via Mersin 10, Turkey

H. Atacag Erkurt (ed.), *Biodegradation of Azo Dyes*,
Hdb Env Chem (2010) 9: 157–167, DOI 10.1007/698_2009_48,

Abbreviations

LiP	Lignin peroxidase
LME	Lignin modifying enzyme
MnP	Manganese peroxidase
PAGE	Polyacrylamide gel electrophoresis
SDS-PAGE	Sodium dodecyl sulphate polyacrylamide gel electrophoresis
SSIIP	Single step isolation and identification procedure
WRF	White rot fungi

1 Introduction

Azo dyes represent the largest group of organic dyes synthesized and account for about 70% of all textile dyes produced. During the dying process most reactive dyes are hydrolysed and later released into waterways. Although these dyes are not toxic by themselves, after release into the aquatic environment, they may be converted into potentially carcinogenic amines [1, 2] that impacted the ecosystem downstream from the mill. The public demands for colour-free discharges to receiving waters have made decolourization of a variety of industrial wastewater a top priority [3]. Microbial decolourization has been claimed to be less expensive and less environmentally intrusive alternative [4]. Many bacteria and fungi are used for the development of biological processes for the treatment of textile effluents [5–7]. Containing various substituents such as nitro and sulfonyl groups, synthetic dyes are not uniformly susceptible to decomposition by activated sludge in a conventional aerobic process. Attempts to develop aerobic bacterial strains for dye decolourization often resulted in a specific strain, which showed a strict ability on a specific dye structure [8]. The use of lignin-degrading white rot fungi (WRF) has attracted increasing scientific attention, as these organisms are able to degrade a wide range of recalcitrant organic compounds. Their lignin modifying enzymes (LME), that is MnP, LiP and laccases, are directly involved in the degradation of not only lignin in their natural lignocellulosic substrates [9, 10] but also various xenobiotic compounds [11, 12] including dyes [13–18]. Peroxidases and laccases of WRF are oxidative enzymes, which do not need any other cellular components to work. They have broad substrate specificity and are able to transform a wide range of toxic compounds. These enzymes, which are widely distributed in nature, have been studied for many years because of their potential use as biocatalysts in pulp and paper bleaching, wastewater treatment, soil remediation, on-site waste destruction and medical diagnostics [19–23].

2 White Rot Fungi Capable of Decolorizing Azo Dyes

List of selected white rot fungi are given in Table 1.

Table 1 Selected white rot fungi and their enzymes able to decolorize azo dyes

WRF	Enzyme	Dye	References
Phanerochaete chrysosporium	LiP	Diazo dyes	[52]
	LiP	Reactive Brilliant Red K-2BP	[53]
	LiP and MnP		
	MnP and β-glucosidase	Amaranth, new coccine, and Orange G	[54]
Trametes vesicolor	–	Reactive Red 2	[55]
	–	Remazol Black B	[56]
Coriolus versicolor	Laccase	Drimarene Blue	[16]
Funalia trogii	Laccase	Astrazone Blue	[34]
	Laccase	Drimarene Blue	[16]
Pleurotus ostreatus	Laccase	Drimarene Blue	[16]
	LiP	Disperse Orange 3	[57]
	–	Methyl Red and Congo Red	
	LiP	Disperse Orange 3	[58]
		Disperse Yellow 3	
Phanerochaete sordida	MnP	Reactive Red 120	[59]
Pleurotus sajorcaju	Laccase	Amaranth, new coccine, and Orange G	[14]
	Laccase	Reactive Black 5	[17]
Irpex lacteus	–	Methyl Red and Congo Red	[60]
		Reactive Orange 16, Congo Red, Reactive Black 5, Naphthol Blue Black, Chicago Sky Blue	
	MnP		[61]
Ganoderma lucidum	Laccase	Reactive Black 5	[42]
Ganoderma sp. WR-1	LiP	Amaranth	[62]
Ischnoderma resinosum	Laccase	Orange G	[15]
Dichomitus squalens	Laccase and MnP	Orange G	[15]
Pleurotus calyptratus	Laccase	Orange G	[15]
Strain L-25 (newly isolated white rot fungus)	MnP	Direct-Orange 26, Direct Red 31, Direct Blue 71, Acid Orange 56, Acid Red 6, Mordant Yellow 3, Mordant Blue 13, Mordant Black 11, Reactive Orange 16, Reactive Black 5	[25]
Lentinula edodes	MnP	Congo Red, Trypan Blue, Amido Black	[13]

3 Enzymes of White Rot Fungi Involved in Azo Dye Decolorization

WRF are key regulators of the global C-cycle. Some WRF produce all three LME, while others produce only one or two of them [10]. The main LME are oxidoreductases, that is two types of peroxidases, LiP and MnP, and a phenoloxidase Laccase. Catalytic cycles of peroxidases and laccases are given in Figs. 1 and 2, respectively. LME are produced by WRF during their secondary metabolism.

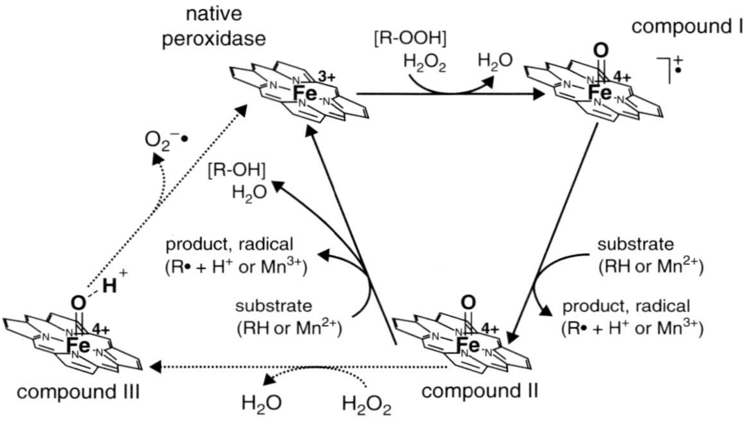

Fig. 1 Generic scheme of the catalytic cycle of peroxidases (taken from [24])

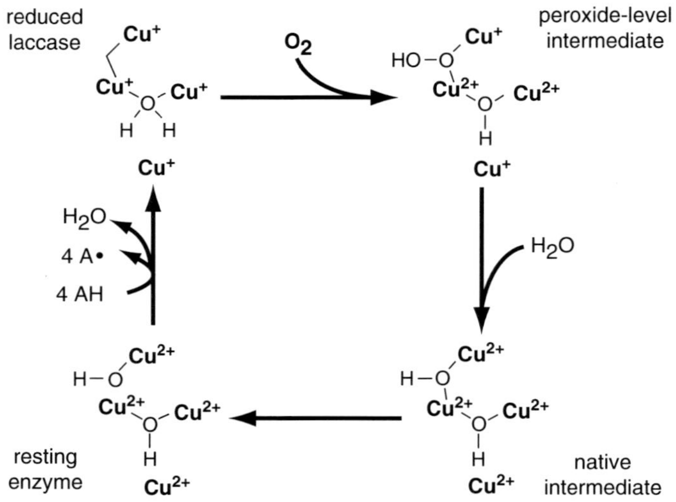

Fig. 2 The catalytic cycle of laccases (taken from [24])

Synthesis and secretion of these enzymes are often induced by limited nutrient (C or N) levels [24].

The proposed mechanism for the functionality of MnP involves the oxidation of manganous ions Mn^{2+} to Mn^{3+}, which is then chelated with organic acids. The chelated Mn^{3+} diffuses freely from the active site of the enzyme and can oxidize secondary substrates [25].

LiP catalyze several oxidations in the side chains of lignin and related compounds [26] by one-electron abstraction to form reactive radicals [27]. Also the cleavage of aromatic ring structures has been reported [28]. The role of LiP in ligninolysis could be the further transformation of lignin fragments, which are initially released by MnP.

Fungal laccases as part of the ligninolytic enzyme system are produced by almost all wood rotting basidiomycetes. This group of *N*-glycosylated extracellular blue oxidases with molecular masses of 60–390 kDa [29, 30] contain four copper atoms in the active site (as Cu^{2+} in the resting enzyme). Laccases catalyze the oxidation of a variety of aromatic hydrogen donors with the concomitant reduction of oxygen to water (Fig. 2). Laccase is an oxidase with a redox potential of 780 mV and can catalyse the oxidation of organic pollutants by reduction of molecular oxygen straightforwardly to water in the absence of hydrogen peroxide or even other secondary metabolites [31]. While anthraquinone was directly oxidized by the laccase, azoic and indigo dyes were not the substrates of laccase, and small molecule metabolites mediated the interaction between the dyes and the enzyme [32].

4 Enzyme Production and Decolorization Methods

Most studies on lignin biodegradation and dye decolourization have been carried out using liquid culture conditions [15]. Homogenized mycelium [16] or pellets [33, 34] of WRF can be used for biodegradation of azo dyes. In batch mode, at the beginning of the decolorization process, adsorption of dye by cells might be observed. However, this color sometimes disappeared when enzymes were released by fungal strains [25]. Liquid media including lignocellulosic substrates are also used for ligninolytic enzyme production [35].

In some researches, solid-state fermentation (SSF) is being used as the media for ligninolytic enzyme production [36]. SSF reflect the natural living conditions (i.e. in wood and other lignocellulosic substrates) of these fungi. SSF is defined as the growth of microorganisms on solid materials in the presence of a small amount of free water [37]. The list of different substrates used for the cultivation of microorganisms on SSF is long, including several agricultural materials, such as wheat bran, wheat straw, sugar cane bagasse and corn cob. The choice of corn cob was due to the low amounts of natural coloured pigments found in this material. The pigments found in other lignocellulosic substrates, such as wheat bran and wheat straw, could interfere in the dye decolorization experiments [13].

SSF containing wheat bran and soybean as a substrate was chosen for the production of ligninolytic enzymes for *Funalia trogii* ATCC200800 [18] as it mimics the natural environment of the WRF and permits the concentration of dyes by absorption process prior to biological treatment [4, 38, 39]. It is possible to stimulate the yield of laccase activity of *Trametes versicolor* by using several agricultural wastes [40].

Crude enzyme of *Earliella scabrosa* obtained in SSF showed higher decolourization percentage of Navy FNB and Red FN-3G dyes than *Trametes maxima* and *Ganoderma zonatum* (B-18). *T. maxima* exhibited the best decolourization percentage in submerged cultures supplemented with Navy FNB, Red FN-3G and yellow P-6GS dyes. Growing biomass of *T. maxima* could supply other enzymes and mediators for dye transformation. Peculiar behaviour was observed with *G. zonatum* (B-18); it had a similar dyes biodegradation in both liquid and solid bed fermentation and there was no positive correlation between ligninolytic enzymes production and decolourization pattern. The employment of crude enzymes produced in the solid bed of bagasse could be an attractive option for biological removal of textile dyes [41].

Forest residue wood chips contain a mixture of fungi and bacteria, which is an advantage when complex molecules should be degraded. The wood chips furthermore provide the microorganisms with carbon source, which make the addition of, for example, glucose unnecessary. The decolourization of a mixture of 200 mg/L each of Reactive Black 5 and Reactive Red 2 dye was studied in batch experiments using microorganisms growing on forest residue wood chips in combination with or without added WRF, *Bjerkandera* sp. BOL 13. The microorganisms growing on the forest residue wood chips decolourized the mixture of the two dyes; adding extra nutrients approximately doubled the decolourization rate [42].

Dye decolorizing potential of the WRF *Ganoderma lucidum* KMK2 was demonstrated for recalcitrant textile dyes. *G. lucidum* produced laccase as the dominant lignolytic enzyme during SSF of wheat bran, a natural lignocellulosic substrate. Crude enzyme shows excellent decolorization activity to anthraquinone dye Remazol Brilliant Blue R without redox mediator, whereas diazo dye Remazol Black-5 (RB-5) requires a redox mediator [43].

Funalia trogii ATTC 200800 pellets and enzymes were used wherein an efficient decolourization was observed within 24 h [16, 34]. The direct decolourization of textile dyes by crude enzymes of *F. trogii* ATCC200800 would provide a cost-effective solution for textile industry. On the other hand, using pellets would also provide a cost-effective solution as repeated addition of dyes is possible. Yesilada et al. reported a 86% decolorization efficiency at the end of tenth cyle [34]. Repeated-batch mode represents a potential alternative mode of fermentation, in which medium or some part of the medium is drawn and fresh medium is refilled periodically without changing the pellets [24]. This process allows the maintenance of long-term activity of the pellet for a long period of time and achieves better results compared with batch cultivation [44]. With this method, it also possible to store the pellets and reuse them. Thus, repeated-batch-type laccase production represents a process which may be applicable for industrial purposes [33].

Using purified enzymes of WRF is another method used for degradation of azo dyes [45]. Purified laccase from *Pleurotus sajorcaju* was reported to be used for decolorization of Reactive Black 5, and increased decolourization was observed with increase in enzyme concentration [43].

Both purified laccase as well as the crude enzyme from the WRF *Cerrena unicolor* were used to convert the dyes in aqueous solution. Biotransformation of the dyes was followed spectrophotometrically and confirmed by high performance liquid chromatography. The results indicate that the decolorization mechanism follows MichaeliseMenten kinetic and that the initial rate of decolorization depends both on the structure of the dye and on the concentration of the dye. Surprisingly, one recalcitrant azo dye (AR 27) was decolorized merely by purified laccase in the absence of any redox mediator [46].

5 Detection of Enzymes Responsible for Azo Dye Decolorization

Measuring Lignin peroxidase, Laccase and MnP activities in decolorization medium is a method to determine the enzyme responsible for decolorization [15, 17, 25].

Lentinula (*Lentinus*) *edodes* produced only Mn peroxidase, and the production of both laccase and lignin peroxidase was, apparently, negligible. Consider that a strict relation between the production of Mn peroxidase and the dye decolorization ability was observed in vivo [13].

Statistical analysis of enzyme amounts could be used to demonstrate which enzyme plays an important role in the decolourization process of azo dyes, and it was reported that the complete decolourization time and enzyme activity are negatively correlative [47].

Molecular masses of the same enzymes of different species are different. Molecular mass of the laccase of *Pleorotus ostreatus* was found to be 66.8 kDa by sodium dodecyl sulphate polyacrylamide gel electrophoresis (SDS-PAGE) [48]. Purified enzyme of *T. versicolor* having a single band with a molecular mass of ~68 kDa was in the same range with the molecular weights of laccase isoforms isolated from 2,5-xylidine-induced cultures of *T. versicolor* [49].

Using SDS-PAGE or native polyacrylamide gel electrophoresis (PAGE) methods is another method to determine enzyme(s) responsible for decolorization. The degradation of the disazo dye Chicago Sky Blue 6B by a purified laccase from *Pycnoporus cinnabarinus* showed a band having a molecular size of 63 kDa determined by SDS-PAGE [50]. Unyayar et al. had reported the Drimarene Blue X3LR decolourizing enzymatic activity in the culture filtrate of *F. trogii* by using SDS-PAGE [18]. In this method, two SDS-PAGEs were performed. One of them was used for determining molecular weight of protein bands (Lane A, Fig. 3). The other one was used for single step isolation and identification procedure. The staining activity was done with Drimarene Blue X3LR dye and guaicol after the gel was re-natured. After the gel was stained with Drimarene Blue X3LR dye and incubated

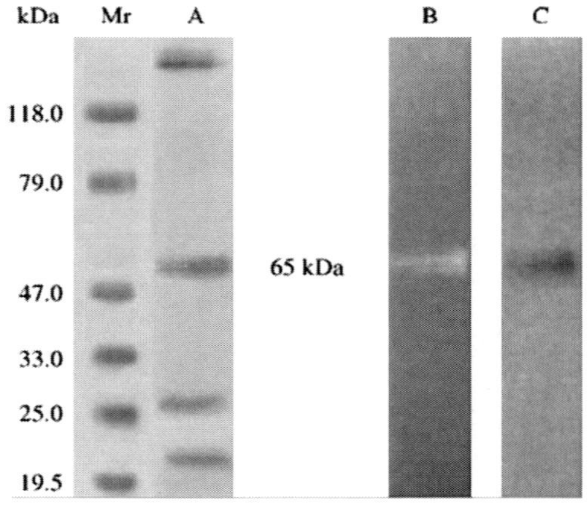

Fig. 3 SDS-PAGE Photograph: Separation (Lane Mr and A) and activity staining (Lane B and C) of the crude filtrate of *Funalia trogii*. Lane Mr standard molecular weight markers (β-galactosidase, 118.0 kDa; bovine serum albumin, 79.0 kDa; ovalbumin, 47.0 kDa; carbonic anhydrase, 33.0 kDa; β-lactoglobulin, 25.0 kDa; and lysozyme, 19.5 kDa). Relative mobilities of the standard markers vs. common logarithms of their molecular masses were plotted.With the linear regression output, the molecular masses of the proteins in the crude filtrate were estimated (taken from [18])

at 30°C for 30 min, a colourless zone was observed (Lane B, Fig. 3). This colourless zone was found to be equal to 65 kDa in Lane A (Fig. 3). Afterwards, the gel was treated with guaicol. This colourless zone turned into orange colour after incubation with guaicol, which is a classical indicator and substrate of laccase (Lane C, Fig. 3), and so it was concluded that this enzyme responsible for decolourization of Drimarene Blue X3LR was a laccase [18].

A similar method was used by Murugesan et al. PAGE of crude enzyme and oxidation of guaicol on gels confirm that the laccase enzyme was the major enzyme involved in the decolorization of RB5. Native and SDS-PAGE indicates the presence of single laccase with molecular weight of 43 kDa [43].

A microtitre plate-based method was developed for a fast screening of numerous fungal strains for their ability to decolourize textile dyes. In 3 days, this method allowed to estimate significant fungal decolourization capability by measuring the absorbance decrease on up to 10 dyes. WRF strains belonging to 76 fungal genera were compared with regards to their capability to decolourize five azo and two anthraquinone dyes as well as the dyes mixture. The most recalcitrant dyes belonged to the azo group. Several new species unstudied in the bioremediation field were found to be able to efficiently decolourize all the dyes tested [51].

Decolorization of azo dyes by WRF technology improvements will require integration of all major areas of industrial biotechnology: novel enzymes and microorganisms, functional genomics, protein engineering, biomaterial development, bioprocess design and applications.

The enzymes of WRF will play a significant role for the working of treatment processes. As a result, the mechanical equipments will be reduced and also pre-investment expenses will drop. The biotechnological methods presented in this work will be expected to reduce the operational cost.

References

1. Chung KT, Stevens SE (1993) Decolourisation of azo dyes by environmental microorganisms and helminthes. Environ Toxicol Chem 12:2121–2132
2. Spadaro JT, Gold MH, Renganadhan V (1992) Degradation of azo dyes by the lignin-degrading fungus Phanerochaete crysosporium. Appl Environ Microbiol 58:2397–2401
3. Hao OJ, Kim H, Chiang PC (2000) Decolourisation of wastewater. Crit Rev Environ Sci Technol 30:449–505
4. Robinson T, McMullan G, Marchant R, Nigam P (2001) Remediation of dyes in textile effluent: a critical review on current treatment technologies with a proposed alternative. Bioresour Technol 77:247–255
5. Churchley JH, Greaves AJ, Hutchings MG, James AE, Phillips DAS (2000) The development of a laboratory method for quantifying the bioelimination of anionic, water-soluble dyes by a biomass. Water Res 34:1673–1679
6. Moreira MT, Feijoo G, Sierra-Alvarez R, Lema J, Field JA (1997) Biobleaching of oxygen delignified kraft pulp by several white rot fungal strains. J Biotechnol 53:237–251
7. Razo-Flores E, Luijten M, Donlon B, Lettinga G, Field J (1997) Biodegradation of selected azo dyes under methanogenic conditions. Water Sci Technol 36:65–72
8. Kulla HG (1981) Aerobic bacterial decolourisation of azo dyes. FEMS Symp 12:387–399
9. Becker HG, Sinitsyn AP (1993) Mn-peroxidase from Pleurotus ostreatus: the action on the lignin. Biotechnol Lett 15:289–294
10. Hatakka A (1994) Lignin-modifying enzymes from selected white-rot fungi: production and role in lignin degradation. FEMS Microbiol Rev 13:125–135
11. Pointing SB (2001) Feasibility of bioremediation by white-rot fungi. Appl Microbiol Biotechnol 57:20–33
12. Scheibner K, Hofrichter M, Fritsche W (1997) Mineralization of 2-amino-4, 6-dinitrotoluene by manganese peroxidase of the white-rot fungus Nematoloma frowardii. Biotechnol Lett 19:835–839
13. Boer CG, Obici L, de Souza CGM, Peralta RM (2004) Decolorization of synthetic dyes by solid state cultures of *Lentinula* (*Lentinus*) *edodes* producing manganese peroxidase as the main ligninolytic enzyme. Bioresour Technol 94:107–112
14. Chagas EP, Durrant LR (2001) Decolorization of azo dyes by *Phanerochaete chrysosporium* and *Pleurotus sajorcaju*. Enzyme Microb Technol 29(8–9):473–477
15. Eichlerova I, Homolka L, Lisa L, Nerud F (2005) Orange G and Remazol Brilliant Blue R decolorization by white rot fungi *Dichomitus squalens* Ischnoderma resinosum and Pleurotus calyptratus. Chemosphere 60:398–404
16. Erkurt EA, Unyayar A, Kumbur H (2007) Decolorization of synthetic dyes by white rot fungi, involving laccase enzyme in the process. Process Biochem 42:1429–1435
17. Murugesan K, Dhamija A, Nam IH, Kim YM, Chang YS (2007) Decolourization of reactive black 5 by laccase: optimization by response surface methodology. Dyes Pigm 75:176–184
18. Unyayar A, Mazmanci MA, Atacag H, Erkurt EA, Coral GA (2005) Drimaren Blue X3LR dye decolorizing enzyme from Funalia trogii: one step isolation and identification. Enzyme Microb Technol 36:10–16
19. Dec J, Bollag JM (1990) Detoxification of substituted phenols by oxidoreductive enzymes through polymerization reactions. Arch Environ Contam Toxicol 19:543–550

20. Duran N, Esposito E (2000) Potential applications of oxidative enzymes and phenoloxidase-like compounds in wastewater and soil treatment: a review. Appl Catal B Environ 28:83–99
21. Gianfreda L, Sannino F, Filazzola MT, Leonowicz A (1998) Catalytic behavior and detoxifying ability of a laccase from the fungal strain Cerrena unicolor. J Mol Catal B Enzym 4:13–23
22. Unyayar A, Demirbilek M, Turkoglu M, Celik A, Mazmanc MA, Erkurt EA, Ünyayar S, Cekic Ö, Atacag H (2006) Evaluation of cytotoxic and mutagenic effects of Coriolus versicolor and Funalia trogii extracts on mammalian cells. Drug Chem Toxicol 1:69–83
23. Van Deurzen MPJ, Seelbach K, van Rantwijk F, Kragl U, Sheldon RA (1997) Chloroperoxidase: use of a hydrogen peroxide-stat for controlling reactions and improving enzyme performance. Biocatal Biotransformation 15:1–16
24. Wesenberg D, Kyriakides I, Agathos N (2003) White-rot fungi and their enzymes for the treatment of industrial dye effluents. Biotechnol Adv 22:161–187
25. Kariminiaae-Hamedaani HR, Sakurai A, Sakakibara M (2007) Decolorization of synthetic dyes by a new manganese peroxidase-producing white rot fungus. Dyes Pigm 72:157–162
26. Tien M, Kirk TK (1983) Lignin-degrading enzyme from the Hymenomycete *Phanerochaete chrysosporium* Burds. Science 221:661–663
27. Kersten PJ, Tien M, Kalyanaraman B, Kirk TK (1985) The ligninase from *Phanerochaete chrysosporium* generates cation radicals from methoxybenzenes. J Biol Chem 260:2609–2612
28. Umezawa T, Higuchi T (1987) Mechanism of aromatic ring cleavage of h-O-4 lignin substructure models by lignin proxidase. FEBS Lett 218:255–260
29. Call HP, Mucke I (1997) Minireview: history, overview and applications of mediated lignolytic systems, especially laccase-mediator-systems (Lignozym-Process). J Biotechnol 53:163–202
30. Reinhammar B (1984) Laccase. In: Lontie R (ed) Copper proteins and copper enzymes, vol 3. CRC, Boca Raton, pp 1–36
31. Thurston CF (1994) The structure and function of fungal laccase. Microbiology 140:19–26
32. Levin L, Forchiassin F (2001) Ligninolytic enzymes of the white rot basidiomycete *Trametes trogii*. Acta Biotechnol 21:179–186
33. Birhanli E, Yesilada O (2006) Increased production of laccase by pellets of *Funalia trogii* ATCC 200800 and *Trametes versicolor* ATCC 200801 in repeated-batch mode. Enzyme Microb Technol 39:1286–1293
34. Yesilada O, Asma D, Cing S (2003) Decolourization of textile dyes by fungal pellets. Process Biochem 38:933–938
35. Pickard MA, Vandertol H, Roman R, Vanzquez-Duhalt R (1999) High production of ligninolytic enzymes from white rot fungi in ceral bran liquid medium. Can J Microbiol 45:627–631
36. Nam IH, Kim YM, Chang YS (2007) Decolorization of reactive dyes by a thermostable laccase produced by Ganoderma lucidum in solid state culture. Enzyme Microb Technol 40:1662–1672
37. Pandey A, Selvakumar P, Soccol CR, Nigam P (1999) Solid-State fermentation for the production of industrial enzymes. Curr Sci 77:149–162
38. Robinson T, Chandran B, Nigam P (2002) Effect of pre-treatments of three waste residues, wheat straw, corncobs and barley husks on dye adsorption. Bioresour Technol 85:119–124
39. Zilly A, de Souza CGM, Barbosa-Tessmann IP, Peralta RM (2002) Decolourisation of industrial dyes by a Brazilian strain of Pleurotus pulmonarius producing laccase as the sole phenol-oxidizing enzyme. Folia Microbiol 47:315–319
40. Lorenzo MD, Moldes D, Rodriguez Couto S, Sanroman A (2002) Improving laccase production by employing different lignocellulosic wastes in submerged cultures of *Trametes versicolor*. Bioresour Technol 82:109–113
41. Guerra G, Domínguez O, Ramos-Leal M, Manzano AM, Sanchez MI, Hernandez I, Palacios J, Arguelles J (2008) Production of laccase and manganese peroxidase by white-rot fungi from sugarcane bagasse in solid bed: use for dyes decolourisation. Sugar Tech 10(3):260–264
42. Forss J, Welander U (2009) Decolourization of reactive azo dyes with microorganisms growing on soft wood chips. Int Biodeterior Biodegradation 63:752–758

43. Murugesan K, Nam IH, Kim YM, Chang YS (2007) Decolorization of reactive dyes by a thermostable laccase produced by Ganoderma lucidum in solid state culture. Enzyme Microb Technol 40:1662–1672
44. Jiang-Ping L, Wei L, Li-Ming X, Pei-Lin C (2001) Production of laccase by Coriolus versicolor and its application in decolorization of dyestuffs. I. Production of laccase by batch and repeated-batch processes. J Environ Sci 15:1–4
45. Soares GMB, Amorim MTP, Hrdina R, Costa-Ferreira M (2002) Studies on the biotransformation of novel disazo dyes by laccase. Process Biochem 37:581–587
46. Michniewicz A, Ledakowicz S, Ullrich R, Hofrichter M (2008) Kinetics of the enzymatic decolorization of textile dyes by laccase from *Cerrena unicolor*. Dyes Pigm 77:295–302
47. Hailei W, Pingb L, Mina P, Zhijuna Z, Guanglib Y, Guoshengb L, Jianminga Y (2009) Rapid decolourization of azo dyes by a new isolated higher manganese peroxidase producer: *Phanerochaete* sp. HSD. Biochem Eng J 46:327–333
48. Hublik G, Schinnera F (2000) Characterization and immobilization of the laccase from Pleurotus ostreatus and its use for the continuous elimination of phenolic pollutants. Enzyme Microb Technol 27:330–336
49. Hofer C, Schlosser D (2000) Novel enzymatic oxidation of Mn+2 to Mn+3 catalyzed by a fungal laccase. FEBS 451:186–190
50. Schliephake K, Mainwaring DE, Lonergan GT, Jones KI, Baker WL (2000) Transformation and degradation of the disazo dye chicago sky blue by a purified laccase from Pycnoporus cinnabarinus. Enzyme Microb Biotechnol 27:100–107
51. Lucas M, Mertens V, Corbisier A-M, Vanhulle S (2008) Synthetic dyes decolourisation by white-rot fungi: development of original microtitre plate method and screening. Enzyme Microb Technol 42:97–106
52. Paszczynski A, Pasti MB, Goszczynski S, Crawford DL, Crawford RL (1991) New approach to improve degradation of recalcitrant azo dyes by *Streptomyces* spp. and *Phanerochaete chrysosporium*. Enzyme Microb Technol 13(5):378–384
53. Adosinda M, Martins M, Ferreira IC, Santos IM, Queiroz MJ, Lima N (2001) Biodegradation of bioaccessible textile azo dyes by *Phanerochaete chrysosporium*. J Biotechnol 89 (2–3):91–98
54. Yu G, Wen X, Li R, Qian Y (2006) In vitro degradation of a reactive azo dye by crude ligninolytic enzymes from nonimmersed liquid culture of *Phanerochaete chrysosporium*. Process Biochem 41(9):1987–1993
55. Nilsson I, Moller A, Mattiasson B, Rubindamayugi MST, Welander U (2006) Decolorization of synthetic and real textile wastewater by the use of white-rot fungi. Enzyme Microb Technol 38:94–100
56. Aksu Z, Kilic NK, Ertugrul S, Donmez G (2007) Inhibitory effects of chromium(VI) and Remazol Black B on chromium(VI) and dyestuff removals by *Trametes versicolor*. Enzyme Microb Technol 40:1167–1174
57. Zhao X, Hardin IR, Hwanga HM (2006) Biodegradation of a model azo disperse dye by the white rot fungus *Pleurotus ostreatus*. Int Biodeterior Biodegradation 57(1):1–6
58. Zhao X, Hardin IR (2007) HPLC and spectrophotometric analysis of biodegradation of azo dyes by *Pleurotus ostreatus*. Dyes Pigm 73:322–325
59. Harazono K, Watanabe Y, Nakamura K (2003) Decolorization of azo dye by the white-rot basidiomycete *Phanerochaete sordida* and by ts manganese peroxidase. J Biosci Bioeng 95(5):455–495
60. Novotny C, Rawal B, Bhatt M, Patel M, Sasek V, Molitoris PH (2001) Capacity of *Irpex lacteus* and *Pleurotus ostreatus* for decolorization of chemically different dyes. J Biotechnol 89(2–3):113–122
61. Novotny C, Svobodova K, Erbanova P, Cajthamal T, Kasinath A, Lang E (2004) Lignolytic fungi in bioremediation: extracellular enzyme production and degradation rate. Soil Biol Biochem 36:1545–1551
62. Revankar MS, Lele SS (2007) Synthetic dye decolourization by white rot fungus, Ganoderma sp. WR-1. Bioresour Technol 98:775–780

Decolorization of Azo Dyes by Immobilized Fungi

Mehmet Ali Mazmanci

Abstract Decolorization has recently become an area of major scientific interest as indicated by the large quantity of related research reports. During the past two decades, several color removal techniques have been reported, few of which have been accepted by some industries. There is a need to find alternative technologies that are effective in decolorizing dyes from large volume of effluents. Alternative technologies such as decolorization with fungi are still in progress. Especially, ligninolytic fungi and their extracellular oxidative enzymes have been reported to be responsible for the decolorization of dyes. Immobilization applications seem to be more encouraging than those with free cells, because it allows using microbial cells and support materials repeatedly. This chapter reviews the widely used immobilization materials and the application of fungal immobilization to dye decolorization process.

Keywords Azo dye, Decolorization, Fungi, Immobilization

Contents

M.A. Mazmanci
Faculty of Engineering, Department of Environmental Engineering, Mersin University, Ciftlikkoy Campus, 33343 Mersin, Turkey
e-mail: mazmanci@mersin.edu.tr

H. Atacag Erkurt (ed.), *Biodegradation of Azo Dyes*,
Hdb Env Chem (2010) 9: 169–181, DOI 10.1007/698_2009_51,

Abbreviations

Ca-ALG	Calcium alginate
CTS	Chitosan
DB15	Direct Blue 15
DMW	Dry mycelium weight
LiP	Lignin peroxidase
MnP	Manganese peroxidase
Na-ALG	Sodium alginate
Na-CMC	Na-carboxymethyl-cellulose
PBR	Packed-bed reactor
PuF	Polyurethane foam
PVA	Polyvinyl alcohol
PW	Pine wood
RB49	Reactive Blue 49
RB5	Reactive Black 5
RBBR	Remazol Brilliant Blue R
RO16	Reactive Orange 16
RR243	Reactive Red 243

1 Introduction

Biological methods are generally considered environmental-friendly, as they can lead to complete mineralization of organic pollutants at low cost. Azo compounds, widely used in a number of industries, are xenobiotic in nature (Fig. 1); only one natural azo compound (4–40 dihydroxy azo benzene) has been reported so far [1]. Thus they can be expected to be recalcitrant to biodegradation. It is generally observed that dyes resist biodegradation in conventional activated sludge treatment units [2]. It is now known that several microorganisms including yeasts, algae, bacteria, and fungi or their enzymes can decolorize and even completely mineralize many azo dyes under certain environmental conditions [2–13].

The role of fungi in the treatment of wastewater has been extensively researched [14–16]. Fungus has proved to be a suitable organism for the treatment of textile effluent and dye removal. Based on the mechanism involved, these studies can be grouped into bioaccumulation, biodegradation, and biosorption. Bioaccumulation is the accumulation of pollutants by actively growing cells by metabolism [17]. Biodegradation is an energy-dependent process and involves the breakdown of dye into various by-products through the action of various enzymes. Fungi can produce the lignin-modifying enzymes, such as laccase, lignin peroxidase (LiP), and manganese peroxidase (MnP), to mineralize and/or to decolorize azo dyes [18–22]. Biosorption is defined as binding of solutes to the biomass by processes that do not

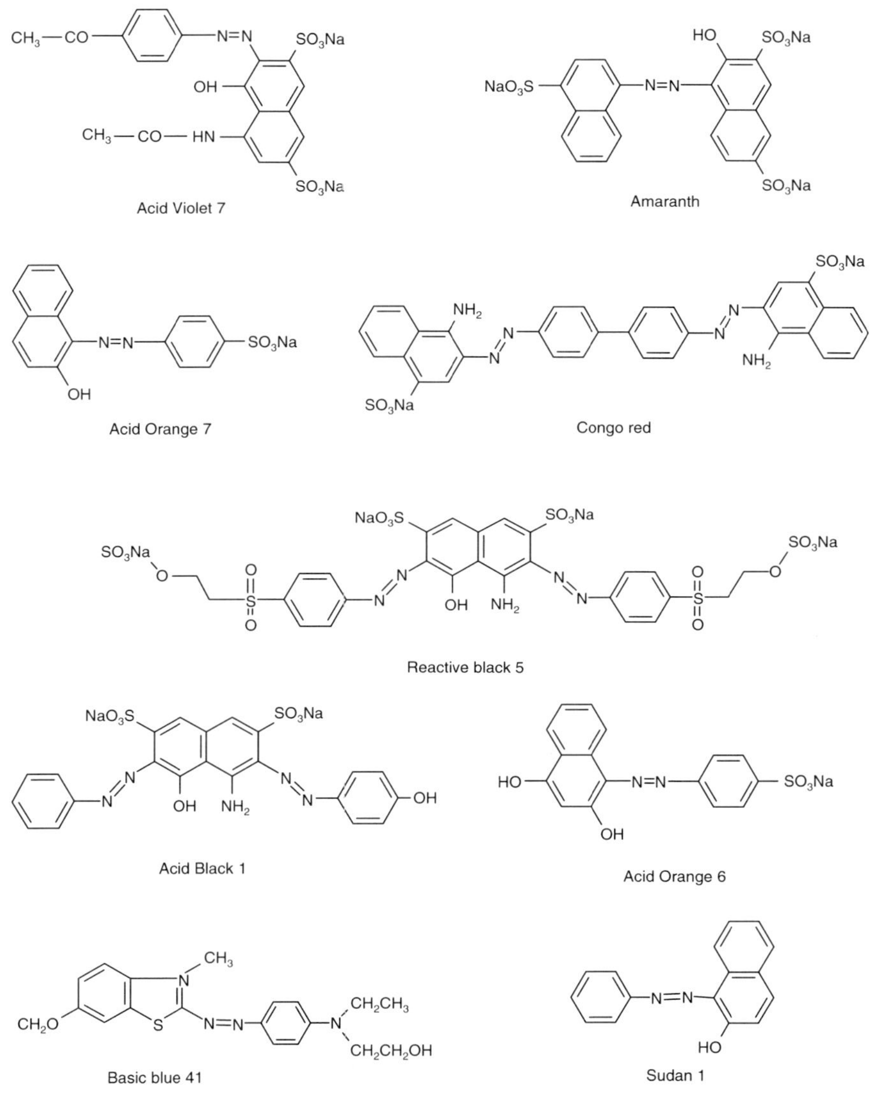

Fig. 1 Chemical structure of synthetic dyes most frequently studied in decolorization experiments by fungal strains

involve metabolic energy or transport, although such processes may occur simultaneously where live biomass is used. Therefore, it can occur in either living or dead biomass [23]. Many genera of fungi have been employed for the dye decolorization either in living or dead form.

Fungal cultures are used as free or immobilized cultures for decolorization processes under static and/or agitated conditions. Free cell cultures could decolorized the dye and/or textile effluent, but it has some operational problems such as shear force, cell stability in agitated conditions. Immobilized fungal cells offer some advantages over free cells, which enhance decolorization efficiency, cell stability, reuse of biomass easier liquid–solid separation, and minimal clogging in continuous-flow systems. Cell immobilization may also protect cells against shear force, toxic compounds, and pH [24–27]. Moreover, cell immobilization is reduced in protease activity and contamination risk [28].

Many reviews have been summarized about decolorization of dyes or colored real effluents [28–34]. Furthermore, many papers have been published dealing with decolorization of different structural dyes by fungi. This chapter is considered about the decolorization of azo dyes by immobilized fungi; reports are on progress.

2 Immobilization

Whole cell immobilization was defined by Karel et al. [35] as, "the physical confinement or localization of intact cells to a certain region of space with preservation of some desired catalytic activity" or by Anderson [36] as, "there is a physical confinement or localization of microorganisms that permits their economic reuse."

Basically, two types of cell immobilization are used: entrapment and attachment. For entrapment technique, natural or synthetic polymers have been used for cell immobilization. In the former, the microorganisms are entrapped in the agar, alginate, chitosan (CTS), cellulose derivatives, or other polymeric matrixes like gelatin, collagen, and polyvinyl alcohol (PVA) [37–39]. In the latter, synthetic foams like polyurethane foam (PuF), nylon sponge, or stainless steel sponge have been used for attachment procedure [40–42]. Natural supports like wheat straw, jute, hemp, maple woodchips, pine wood (PW), and *Luffa cylindrica* sponge have also been used to immobilize fungi [43, 44]. These materials mimic what occurs naturally when cells grow on surfaces or within natural structures. Thus, they can provide them with additional nutrient and stimulate the production of ligninolytic enzymes [44, 45].

3 Dye Decolorization by Immobilized Fungi

Table 1 summarizes the azo dye decolorization by fungi immobilized on different supports.

Cell immobilized cultures of *Phanerochaete chrysosporium* and repeated-batch decolorization were reported by Yang and Yu [46]. Diazo-dye Red 533 was decolorized by PuF immobilized culture, and decolorization efficiency of 80% or higher was achieved within a period of 1 or 2 days.

Table 1 Azo dye decolorization by fungi immobilized on different supports

Azo dye	Fungus	Support material	References
Acid Black 52	*F. trogii*	Na-ALG beads	[55]
Diazo dye Red 533	*P. chrysosporium*	PuF	[46]
Poly S-119	*C. lignorum* CL1	Plastic packing	[47]
Orange II	Unidentified fungus F29	Na-ALG beads	[48]
	C. versicolor RC3	PuF	[65]
	P. chrysosporium	PuF	[50]
Acid Violet 7	*T. versicolor*	Activated carbon powder	[49]
Amaranth	*T. versicolor* ATCC 20869	Wheat straw, jute, hemp, maple woodchips, nylon, polyethylene teraphthalate fibers	[44]
RBBR	*I. lacteus*	PuF, PW	[51]
Astrazon Red dye	*F. trogii*	Activated carbon	[52]
Drimarene Blue	*F. trogii*	*L. cylindrica* sponge	[45]
Reactive Black 5	*F. trogii*	*L. cylindrica* sponge	[43]
	P. chrysosporium	PuF	[57]
	T. pubescens	Stainless steel sponges	[66]
Acid Orange	*P. chrysosporium*	Alginate beads	[53]
Acid Red 114	*P. chrysosporium*	Alginate beads	[53]
Congo Red	*P. chrysosporium*	Ca-ALG beads	[53]
Direct Yellow 12	*P. chrysosporium*	$ZrOCl_2$-activated pumice	[54]
Direct Green 6			
Direct Brown 2			
Direct Black 38			
Direct Blue 15			
Direct Red 23			
Congo Red			
Direct Orange 26			
Tartrazine			
Acid Black 1	*P. chrysosporium*	PuF	[57]
Reactive Orange 16	*P. chrysosporium*	PuF	[57]
	I. lacteus	PuF	[62]
	D. squalens	PuF, PW	[60]
Basic Blue 41	*P. chrysosporium*	PuF	[57]
Reactive Red 2	*Bjerkandera* sp. strain BOL 13	Birch wood	[56]
Reactive dye K-2BP	*P. chrysosporium*	Nylon nets, PuF	[58]
	A. fumigatus	Na-CMC, Na-ALG, PVA, CTS	[61]
	P. chrysosporium	PuF, stainless steel net, polyamide fiber, fiber glass net	[64]
Reactive Red 243	*T. pubescens* and *P. ostreatus*	PuF	[63]

Decolorization of polymeric dyes Poly R-478 (polyanthraquinone-based) and Poly S-119 (azo dye) by immobilized white rot fungus *Crysosporium lignorum* CL1 on circular plastic packing material in 2L air-lift fermenter was studied by Buckley and Dobson [47]. They also examined the relationship between polymeric dye decolorization and the production of LiP and MnP activity in its statistically growth cultures. The fermenter with modified Kirks medium both with and without addition of $MnSO_4$ was setup.

The dye Poly R-478 was decolorized to a much greater extent and at slightly faster rate when the culture was supplemented with Mn(II), while the opposite was obtained for Poly S-119. They found a correlation between polymeric dye decolorization and peroxidative activity of fungus under static or immobilized condition in air-lift bioreactor. Immobilized culture produced LiP and MnP enzymes over a longer time than static cultures.

Decolorization of azo dye Orange II with unidentified fungus F29 in fedbatch fluidized-bed bioreactor was investigated by Zhang et al. [48]. The decolorization rates of immobilized cell into Na-ALG beats (40–45 mg/L h) were higher compared with the results of a similar experiment with free cells (30–40 mg/L h). Immobilized mycelia were reused continually for Orange II decolorization for more than 2 months.

Complex mycelium pellets of *Trametes versicolor* with activated carbon powder were investigated for decolorization of Acid Violet 7 [49]. The complex pellets showed the best dye removal. The dye was almost completely removed in 6 h. For complex pellets, maximum decolorization rate (V_{max}; mg/L h) and half velocity concentration (K_s; mg/L) was calculated 130.5 and 345.0 in batch system, respectively.

The decolorization of Orange II by immobilized *P. chrysosporium* in a continuous packed-bed reactor (PBR) was investigated [50]. Nearly complete decolorization (95%) with immobilized fungus on PuF was obtained when working at optimal conditions [dye load rate of 0.2 g/l/d, temperature of 37°C, a hydraulic retention time (HRT) of 24 h], and also oxygen gas in a pulsed flow was applied. A correlation between residual MnP activity and decolorization was observed, but no laccase and LiP enzyme activities were detected.

Wheat straw, jute, hemp, maple woodchips, and nylon and polyethylene teraphthalate fibers were tested for surface immobilization and decolorization of Amaranth by *T. versicolor* ATCC 20869 [44]. They found that fungus immobilized on jute, straw, and hemp decolorized amaranth without glucose being added. Decolorization efficiency increased when 1 g/L glucose was added.

Comparison of dye degradation capacities of submerged and stationary liquid cultures and fungal cultures immobilized on PuF or PW cubes was investigated [51]. They found that stationary cultures exhibited higher levels of LiP, MnP, and laccase than submerged cultures and selective inhibitor analysis brought evidence that MnP played a major role in the decolorization of Remazol Brilliant Blue R (RBBR) by *Irpex lacteus*. However, no LiP was detected in PuF or PW immobilized culture. The immobilized cultures of *I. lacteus* also exhibited good capacities for decolorization of industrial effluents containing dyes in mixtures with other technologically important additives.

The decolorization of mono-azo textile dye Astrazon Red dye by free pellets and immobilized on activated carbon of *Funalia trogii* ATCC200800 was studied [52]. The decolorization efficiency of the immobilized pellets after 10 days of operation was found higher (88%) compared with the results of a similar experiment with free

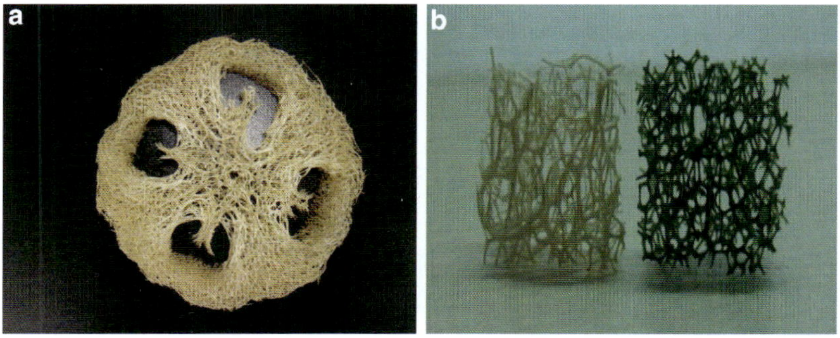

Fig. 2 (**a**) A tropical member of the cucumber family, *Luffa cylindrical*. (**b**) Piece of *L. cylindrica* (*on left*) and PuF (*on right*)

pellets (69%). *F. trogii* pellets rapidly decolorized the dye in 24 h without any visual sorption of any dye to the pellets. They also tested glucose and cheese whey and different concentrations of NH_4Cl as a nitrogen source. Although, no positive effect of nitrogen sources on decolorization performance was detected, decolorization performance of the free pellets remained high and stable in cheese whey-supplemented cultures.

Decolorization of Drimarene Blue K2RL by white rot fungus *F. trogii* was studied by Ayten et al. [45]. Fungus was immobilized on a natural support *L. cylindrica* sponge (Fig. 2). Dye was decolorized by immobilized fungus without adding carbon and nitrogen sources. Maximal decolorization rate (V_{max}) and affinity (K_m) were found to be 12.36 mg dye/L and 193.05 mg dye/L h, respectively. Same support materials and fungus were used to decolorize RB5 by Mazmanci and Ünyayar [43]. They reported that immobilized culture of *F. trogii* decolorized the RB 5 effectively.

Decolorization rate of a 3-day-old culture was found higher (8.22 mg dye/g dmw·day) than others. Maximum dye decolorization was found to be 773.46 mg/L dye after 17 h. They also found that dye decolorization was only due to fungal enzymes (Fig. 3).

The decolorization potential of immobilized *P. chrysosporium* MTCC 787 for azo dyes Acid Orange, Acid Red 114, triphenylmethane dye Methyl Violet, diazoic dye Congo Red, vat dye Vat Magenta, thiazine dye Methylene Blue, and anthraquinone Acid Green was demonstrated by Radha et al. [53]. Decolorization experiments were carried out with immobilized calcium alginate (Ca-ALG) beads of different sizes (2–6 mm).

They found the percentage decolorization decrease with increasing bead diameter for all dyes. Adsorption was determined by Ca-ALG beads (without immobilization) and it showed an initial reduction of 20% of the color. The immobilized fungus in Ca-ALG beads showed a low K_{dye} value for the Congo Red, a high K_{dye} value for Acid Orange and almost a constant value for Acid Red 114. They reported that *P. chrysosporium* was not able to decolorize Acid Green at a concentration

Fig. 3 Before (*on left*) and after (*on right*) decolorization of RB5 by *Funalia trogii* immobilized on *L. cylindrica* sponge

greater than 0.08 g/L. Maximum decolorization for all dyes was found to be more than 75% at the optimum conditions (35°C, pH 4–5, 1.6×10^5 cell/mL). They showed that MnP and LiP were the key enzymes responsible for the decolorization process.

In vitro and in vivo decolorization of structurally different nine direct azo dyes [Chrysophenine (Direct Yellow 12), Direct Green 6, Direct Brown 2, Direct Black 38, Direct Blue 15 (DB15), Congo Red, Direct Orange 26, Tartrazine (Acid Yellow 23), Direct Red 23] by *P. chrysosporium* BKM-F-1767 (ATCC 24725) immobilized on $ZrOCl_2$-activated pumice was studied by Pazarlioglu et al. [54]. A small-scale PBR was operated for decolorization of DB15, which was determined as the best decolorized dye. Repeated batches were found to 95–100%. In this decolorization process, it was observed that MnP played an important role, while there was no obvious role for LiP, and adsorbtion was determined to be a minor mechanism for DB15 decolorization.

The immobilization of the white rot fungus *F. trogii* in Na-ALG beads allowed the decolorization of the dye Acid Black 52 in a stirred tank reactor operated in batch [55]. Three enzymes, laccase, MnP, LiP, secreted by fungus were reported during decolorization process. Results showed that laccase enzyme activity increased with increasing alginate concentration from 0 to 4%. Cell growth at immobilized cultivation was maintained more stably than suspended cultivation. Total amount of removed dye was reported to be 469 mg/L for immobilized cultures and 440 mg/L for suspended cultures.

The decolorization of Reactive Red 2 (azo dye) and Reactive Blue 4 by immobilized fungus *Bjerkandera* sp. strain BOL 13 was studied [56]. Birch wood was used as a carrier material and circular disks were used in the continuous rotating biological contactor. The experiment results showed the fungus to be able to decolorize mixtures of both dyes efficiently. Decolorization was found to be approximately same at 50 and 100 mg/L of dye stuff (96 and 94%, respectively). When the concentration was increased to 200 mg/L, decolorization decreased to 81%.

Continuous culture of immobilized *P. chrysosporium* on PuF was studied for decolorization of 4 different azo dyes [57]. Acid Black 1, Basic Blue 41, Reactive Black 5, and Reactive Orange 16 (RO16) were effectively decolorized depending on the dye concentration.

P. chrysosporium, immobilized on nylon PuF and nets, were studied to decolorize reactive brilliant red K-2BP under nonsterile conditions by Gao et al. [58]. The fungi immobilized on PuF and nylon nets decolorized the azo dye by 52 and 95%, respectively. The system with nylon nets were contaminated easily with yeast, which decreased the decolorization efficiency. Structure of PuF reported that it was benefit to fungal growth in spreading mycelia.

Decolorization of azo dye RO16 by immobilized cultures of *I. lacteus* was compared in three different reactor systems [59]. Different size of PuF was used for immobilization in reactors. Biomass concentration was reported to be 11.6, 8.3, and 4.9 g dw/L in Small Trickle Bed Reactor (STBR), Large Trickle Bed Reactor, and Rotating Disk Bioreactor, respectively. Decolorization rate was found high in STBR, where 90% decolorization rates were achieved after 3 days. Dye decolorization was highly efficient, but no direct relationship between the extracellular enzyme activities (laccase and MnP) and dye decolorization capacity was found.

Šušla et al. [60] investigated RO16 and RBBR decolorization capacity of immobilized *Dichomitus squalens* on PuF and PW in a fixed-bed reactor. Fungus immobilized on PW and PuF decolorized 42 and 73% of azo dye RO16 within 24 h. Similar sorption capacities were reported for support materials (approximately 30–35%). The culture filtrate containing ligninolytic enzymes secreted by fungi to media during incubation were also studied. The culture filtrate containing laccase, MnP, and MIP decolorized 12% of RO16.

Adsorption of reactive dye K-2BP by immobilized *Aspergillus fumigatus* in Na-carboxymethyl-cellulose (Na-CMC), Na-ALG, PVA, and CTS was studied by Wang et al. [61]. The dye culture mediums were almost completely decolorized 48 h using CTS and Na-CMC immobilized beads. The adsorption efficiency of SA and PVA-SA immobilized beads exceeded 92 and 79.8% in 48 h, respectively.

White rot fungus *I. lacteus* immobilized on PuF was studied for finding out of degradation product of RO16 [62]. Dye decolorization reached 80% within 24 h. They suggested that the dye decolorization process in fungal cultures also involved sorption of the dye due to decrease of 10% in absorbance in un-inoculated controls.

Immobilized *Trametes pubescens* MUT 2295 and *Pleurotus ostreatus* MUT 2976 on PuF in bioreactor were studied by Casieri et al. [63] to decolorize Reactive Red 243 (an azo dye, RR243), Reactive Blue 49 (RB49), and RBBR (antraquinone dyes). Low-nitrogen mineral medium was subsequently used for dye decolorization cycles. Both fungi were able to decolorize RB49 and RBBR dyes even at the high concentration (1,000–2,000 ppm), while RR243 was decolorized to a less extent. Decolorization efficiency of *T. pubescens* was reported higher (65%) than that of *P. ostreatus* (45%) for azo dye RR243. Significantly increased laccase enzyme activities were reported for *P. ostreatus* when the industrial dyes were added at 2,000 ppm. Ecotoxicity tests were applied to measure the toxicity of dyes after decolorization. A significant reduction of toxicity was observed, but samples after

T. pubecens treatment presented a lower growth inhibition than *P. ostreatus*-treated samples.

Gao et al. [64] developed a treatment approach by using immobilized white rot fungus *P. chrysosporium* to degrade reactive dye K-2BP. The fungus was immobilized on PuF, stainless steel net, polyamide fiber, fiber glass net and then used for decolorization under sterile and nonsterile conditions. They found that immobilized cultures on PuF had high enzymatic activity (683 U/L for MnP), high decolorization efficiency (69% in 1 day and 93.5% in 3 days), and shorter decolorization period (3 days) under nonsterile conditions. No difference was found under nonsterile and sterile conditions for degradation of dye with the immobilized fungal cultures.

Orange II decolorization by immobilized thermotolerant fungus *Coriolus versicolor* RC3 was investigated [65]. They found that dye decolorization by immobilized fungus on the PuF with 1 g/L of glucose and 0.2 g/L of ammonium oxalate provided a faster decolorization rate. From their results, 1.5 cm^3 of the PuF showed the most suitable size for immobilization when compare with 1.0 and 2.0 cm^3 material size. Orange II decolorization efficiency was enhanced when HRT of the system was increased.

The decolorization of diazo dye RB5 by *T. pubescens* immobilized on stainless steel sponges in a fixed-bed reactor was studied [66]. Laccase production in the presence of RB5 reached its maximum value of 1,025 U/L. They found that decolorization was due to dye adsorption onto the fungus mycelium and dye decolorization by laccase enzymes produced by the fungus.

References

1. Gill M, Strauch RJ (1984) Constituents of *Agaricus xanthodermus* Genevier: the first naturally endogenous azo compound and toxic phenolic metabolites. Z Naturforsch C 39:1027–1029
2. Stolz A (2001) Basic and applied aspects in the microbial degradation of azo dyes. Appl Microbiol Biotechnol 56:69–80
3. Banat IM, Nigam P, Singh D, Marchant R (1996) Microbial decolourization of textile dyes containing effluents: a review. Bioresour Technol 58:217–227
4. Delée W, O'Neill C, Hawkes FR, Pinheiro HM (1998) Anaerobic treatment of textile effluents: a review. J Chem Technol Biotechnol 73:323–335
5. Deveci T, Unyayar A, Mazmanci MA (2004) Production of Remazol Brilliant Blue R decolourising oxygenase from the culture filtrate of *Funalia trogii* ATCC 200800. J Mol Catal B Enzym 30:25–32
6. Kandelbauer A, Erlacher A, Cavaco-Paulo A, Guebitz GM (2004) Laccase catalyzed decolorization of the synthetic dye Diamond Black PV 200 and some structurally related derivatives. Biocatal Biotransformation 22:331–339
7. Mazmanci MA, Unyayar A, Ekiz HI (2002) Decolorization of Methylene Blue by White Rot Fungus Coriolus versicolor. Fresenius' Environ Bull 11:5
8. McMullan G, Meehan C, Conneely A, Kirby N, Robinson T, Nigam P, Banat IM, Marchant R, Smyth WF (2001) Microbial decolourisation and degradation of textile dyes. Appl Microbiol Biotechnol 56:81–87

9. O'Neill C, Hawkes FR, Lourenco ND, Pinheiro HM, Delee W (1999) Color in textile effluents-source, measurement, discharge contents and simulation: a review. J Chem Technol Biotechnol 74:1009–1018
10. Pandey A, Singh P, Iyengar L (2007) Bacterial decolorization and degradation of azo dyes. Int Biodeterior Biodegradation 59:73–84
11. Rai H, Bhattacharyya M, Singh J, Bansal TK, Vats P, Banerjee UC (2005) Removal of dyes from the effluent of textile and dyestuff manufacturing industry: a review of emerging techniques with reference to biological treatment. Crit Rev Environ Sci Technol 35:219–238
12. van der Zee FP, Villaverde S (2005) Combined anaerobic–aerobic treatment of azo dyes – a short review of bioreactor studies. Water Res 39:1425–1440
13. Vandevivere PC, Bianchi R, Verstraete W (1998) Review: treatment and reuse of wastewater from the textile wet-processing industry: review of emerging technologies. J Chem Technol Biotechnol 72:289–302
14. Azmi W, Sani RK, Banerjee UC (1998) Biodegradation of triphenylmethane dyes. Enzyme Microb Technol 22:185–191
15. Brar SK, Verma M, Surampalli RY, Misra K, Tyagi RD, Meunier N, Blais JF (2006) Bioremediation of hazardous wastes – a review. Prac Period Hazard Tox Radioactive Waste Manag 10:59–72
16. Coulibaly L, Agathos SN (2003) Transformation kinetics of mixed polymeric substrates under transitory conditions by Aspergillus niger. Afr J Biotechnol 2:438–443
17. Aksu Z, Dönmez G (2005) Combined effects of molasses sucrose and reactive dye on the growth and dye bioaccumulation properties of *Candida tropicalis*. Process Biochem 40: 2443–2454
18. Chagas EP, Durrant LR (2001) Decolorization of azo dyes by *Phanerochaete chrysosporium* and *Pleurotus sajor-caju*. Enzyme Microb Technol 29:473–477
19. Eichlerova I, Homolka L, Nerud F (2006) Ability of industrial dyes decolorization and ligninolytic enzymes production by different Pleurotus species with special attention on *Pleurotus calyptratus*, strain CCBAS 461. Process Biochem 41:941–946
20. Levin L, Papinutti L, Forchiassin F (2004) Evaluation of Argentinean white rot fungi for their ability to produce lignin-modifying enzymes and decolorize industrial dyes. Bioresour Technol 94:169–176
21. Ozsoy HD, Unyayar A, Mazmanci MA (2005) Decolourisation of reactive textile dyes Drimarene Blue X3LR and Remazol Brilliant Blue R by *Funalia trogii* ATCC 200800. Biodegradation 16:195–204
22. Unyayar A, Mazmanci MA, Atacag H, Erkurt EA, Coral G (2005) A Drimaren Blue X3LR dye decolorizing enzyme from *Funalia trogii* one step isolation and identification. Enzyme Microb Technol 36:10–16
23. Tobin JM, White C, Gadd GM (1994) Metal accumulation by fungi: applications in environmental biotechnology. J Ind Microbiol Biotechnol 13:126–130
24. Abraham TE, Jamuna R, Bansilal CV, Ramakrishna SV (1991) Continuous synthesis of glucoamylase by immobilized fungal mycelium of *Aspergillus niger*. Starch-Starke 43:113–116
25. Fiedurek J, Ilczuk Z (1991) Glucose oxidase biosynthesis using immobilized mycelium of *Aspergillus niger*. World J Microbiol Biotechnol 7:379–384
26. Margaritis A, Merchant FJA (1984) Advances in ethanol production using immobilized cell systems. Crit Rev Biotechnol 2:339–393
27. Vassilev N, Vassileva M (1992) Production of organic acids by immobilized filamentous fungi. Mycol Res 96:563–570
28. Couto SR (2009) Dye removal by immobilised fungi. Biotechnol Adv 27:227–235
29. Anjaneyulu Y, Sreedhara Chary N, Samuel Suman Raj D (2005) Decolourization of industrial effluents-available methods and emerging technologies – a review. Rev Environ Sci Biotechnol 4:245–273
30. Fu Y, Viraraghavan T (2001) Fungal decolorization of dyes: a review. Bioresour Technol 79:251–262

31. Husain Q (2006) Potential applications of the oxidoreductive enzymes in the decolorization and detoxification of textile and other synthetic dyes from polluted water: a review. Crit Rev Biotechnol 26:201–221
32. Kaushik P, Malik A (2009) Fungal dye decolourization: recent advances and future potential. Environ Int 35:127–141
33. Robinson T, McMullan G, Marchant R, Nigam P (2001) Remediation of dyes in textile effluents: a critical review on current treatment technologies with a proposed alternative. Bioresour Technol 77:247–255
34. Wesenberg D, Kyriakides I, Agathos SN (2003) White-rot fungi and their enzymes for the treatment of industrial dye effluents. Biotechnol Adv 22:161–187
35. Karel SF, Libicki SB, Robertson CR (1985) The immobilization of whole cells-engineering principles. Chem Eng Sci 40:1321–1354
36. Anderson JG (1975) Immobilized cell and film reactor system for filamentous fungi. In: Smith JE, Berry DR, Kristiansen B (eds) Fungal technology. The filamentous fungi. Edward Arnold, London, pp 146–170
37. Katzbauer B, Narodoslawsky B, Moser A (1995) Classification system for immobilization techniques. Bioprocess Eng 12:173–179
38. Norton S, D'Amore T (1994) Physiological effects of yeast cell immobilization: applications for brewing. Enzyme Microb Technol 16:365–375
39. Park JK, Chang HN (2000) Microencapsulation of microbial cells. Biotechnol Adv 18:303–319
40. Couto SR, Sanromán MÁ, Hofer D, Gübitz GM (2004) Stainless steel sponge: a novel carrier for the immobilization of the white-rot fungus *Trametes hirsuta* for decolorization of textile dyes. Bioresour Technol 95:67–72
41. Haapala A, Linko S (1993) Production of *Phanerochaete chrysosporium* lignin peroxidase under various culture conditions. Appl Microbiol Biotechnol 40:494–498
42. Nakamura Y, Sawada T, Sungusi MG, Kobayashi F, Kuwahara M, Ito H (1997) Lignin peroxidase production by *Phanerochaete chrysosporium*. J Chem Eng Jpn 30:1–6
43. Mazmanci MA, Ünyayar A (2005) Decolourisation of Reactive Black 5 by Funalia trogii immobilised on Luffa cylindrica sponge. Process Biochem 4:337–342
44. Shin M, Nguyen T, Ramsay J (2002) Evaluation of support materials for the surface immobilization and decoloration of amaranth by Trametes versicolor. Appl Microbiol Biotechnol 60:218–223
45. Ayten N, Ay T, Ünyayar A, Erkurt EA, Mazmanci MA (2005) Decolorization of Drimarene Blue by *Funalia trogii* ATCC200800 immobilized on *Luffa cylindrica* sponge. VI. In: National Environmental Engineering Congress, Mersin, Turkey, 24–26 Oct 2005, Istanbul
46. Yang F, Yu J (1996) Development of a bioreactor system using an immobilized white rot fungus. Part II. Continuous decolorization tests. Bioprocess Eng 16:9–11
47. Buckley KF, Dobson ADW (1998) Extracellular ligninolytic enzyme production and polymeric dye decolourization in immobilized cultures of *Chrysosporium lignorum* CL1. Biotechnol Lett 20:301–306
48. Zhang F, Knapp JS, Tapley KN (1999) Development of bioreactor systems for decolorization of Orange II using white rot fungus. Enzyme Microb Technol 24:48–53
49. Zhang FM, Yu J (2000) Decolourisation of acid violet 7 with complex pellets of white rot fungus and activated carbon. Bioprocess Eng 23:295–301
50. Mielgo I, Moreira MT, Feijoo G, Lema JM (2001) A packed-bed fungal bioreactor for the continuous decolourisation of azo-dyes (Orange II). J Biotechnol 89:99–106
51. Kasinath A, Novotný C, Svobodová K, Patel KC, Šašek V (2003) Decolorization of synthetic dyes by *Irpex lacteus* in liquid cultures and packed-bed bioreactor. Enzyme Microb Technol 32:167–173
52. Cing S, Yesilada O (2004) Astrazon Red dye decolorization by growing cells and pellets of *Funalia trogii*. J Basic Microbiol 44:263–269
53. Radha KV, Regupathi I, Arunagiri A, Murugesan T (2005) Decolorization studies of synthetic dyes using *Phanerochaete chrysosporium* and their kinetics. Process Biochem 40:3337–3345

54. Pazarlioglu NK, Urek RO, Ergun F (2005) Biodecolourization of Direct Blue 15 by immobilized *Phanerochaete chrysosporium*. Process Biochem 40:1923–1929
55. Park C, Lee B, Han EJ, Lee J, Kim S (2006) Decolorization of Acid Black 52 by fungal immobilization. Enzyme Microb Technol 39:371–374
56. Axelsson J, Nilsson U, Terrazas E, Alvarez Aliaga T, Welander U (2006) Decolorization of the textile dyes Reactive Red 2 and Reactive Blue 4 using Bjerkandera sp. Strain BOL 13 in a continuous rotating biological contactor reactor. Enzyme Microb Technol 39:32–37
57. Urra J, Sepulveda L, Contreras E, Palma C (2006) Screening of static culture and comparison of batch and continuous culture for the textile dye biological decolorization by *Phanerochaete chrysosporium*. Brazilian J Chem Eng 23:281–290
58. Gao D, Wen X, Zeng Y, Qian Y (2006) Decolourization of a textile-reactive dye with *Phanerochaete chrysosporium* incubated in different ways under non-sterile conditions. Water Pract Technol 1:10801–10806
59. Tavčar M, Svobodová K, Kuplenk J, Novotný Č, Pavko A (2006) Biodegradation of azo dye RO16 in different reactors by immobilized *Irpex lacteus*. Acta Chim Slov 53:338–343
60. Šušla M, Novotný Č, Svobodová K (2007) The implication of *Dichomitus squalens* laccase isoenzymes in dye decolorization by immobilized fungal cultures. Bioresour Technol 98:2109–2115
61. Wang BE, Hu YY, Xie L, Peng K (2008) Biosorption behavior of azo dye by inactive CMC immobilized *Aspergillus fumigatus* beads. Bioresour Technol 99:794–800
62. Svobodová K, Senholdt M, Novotný C, Rehorek A (2007) Mechanism of Reactive Orange 16 degradation with the white rot fungus *Irpex lacteus*. Process Biochem 42:1279–1284
63. Casieri L, Varese GC, Anastasi A, Prigione V, Svobodová K, Marchisio VF, Novotný Č (2008) Decolorization and detoxication of reactive industrial dyes by immobilized fungi Trametes pubescens and *Pleurotus ostreatus*. Folia Microbiol 53:44–52
64. Gao D, Zeng Y, Wen X, Qian Y (2008) Competition strategies for the incubation of white rot fungi under non-sterile conditions. Process Biochem 43:937–944
65. Srikanlayanukul M, Kitwechkun W, Watanabe T, Khanongnuch C (2008) Decolorization of Orange II by immobilized Thermotolerant white rot fungus *Coriolus versicolor* RC3 in Packed-Bed bioreactor. Biotechnology 7:280–286
66. Enayatzamir K, Alikhani HA, Couto SR (2009) Simultaneous production of laccase and decolorization of the diazo dye Reactive Black 5 in a fixed bad reactor. J Hazard Mater 164:296–300

Decolorization of Azo Dyes by Yeasts

Albino A. Dias, Marco S. Lucas, Ana Sampaio, José A. Peres, and Rui M.F. Bezerra

Abstract Azo dyes are synthetically produced organic molecules and represent the largest group of commercial dyes. Industrial use for coloring purposes generates huge volumes of dyed effluents, which are of environmental concern. Color removal has been achieved by using microorganisms such as filamentous fungi, especially white rot fungi, and bacterial species. In this chapter, we look for a still largely unexplored microbial group – the yeasts, and based on the review of current state of the art, we discuss the potential biotechnological applications in the field of azo dyes bioremediation.

Keywords Azo dyes, Bioremediation, Decolorization, Wastewater, Yeasts

Contents

A.A. Dias (✉), A. Sampaio, and R.M.F. Bezerra
CITAB, UTAD – Universidade de Trás-os-Montes e Alto Douro, Apartado 1013, 5001-801 Vila Real, Portugal
e-mail: jdias@utad.pt

M.S. Lucas and J.A. Peres
Centro de Química, UTAD – Universidade de Trás-os-Montes e Alto Douro, Apartado 1013, 5001-801 Vila Real, Portugal

H. Atacag Erkurt (ed.), *Biodegradation of Azo Dyes*,
Hdb Env Chem (2010) 9: 183–193, DOI 10.1007/698_2009_49,

1 Introduction

Biological color removal of dyed effluents containing azo dyes is not an easy task, mainly because they are synthetically produced xenobiotic compounds characterized by the presence of at least one chromophoric group, the azo bond which, in turn, is also linked to two carbon atoms of neighboring aryl (or heteroaryl) or alkyl derivatives. However, effective decolorization including the breakdown of the azo bond has been achieved by using several microorganisms, notably white rot fungi, other filamentous fungi, and bacterial species [1, 2]. Because the growth rates of filamentous fungi (molds) are usually slow when compared to most yeast species, they present an advantage from a biotechnological point of view (significantly shorter reaction times, which means cost savings). Additionally, like molds, yeasts are resilient microorganisms. They are able to resist unfavorable environments such as low pH, high salt concentration, and high-strength organic wastewaters such as the case of textile effluents. This work looks for the biotechnological potential, evaluated in terms of biodegradation and color removal ability, of a still largely unexplored microbial group in the field of azo dyes bioremediation: the yeasts.

2 Mechanisms of Yeast-Mediated Azo Dyes Decolorization

2.1 Yeast Definition

Although all yeast species form individual cells, they vary in their capacity to produce pseudo-hyphae, for example, *Candida* sp., and true mycelium, for example, *Geotrichum* sp. [3, 4], which can make difficult the distinction between yeasts and molds in some species. In fact, the usual terms "yeast" and "molds" do not have any taxonomic significance. In this chapter we will name yeast according to the definition given by [4]: yeasts, whether ascomycetes or basidiomycetes, are fungi with vegetative growth by budding or fission, never presenting its sexual phase within or upon a fruiting body.

2.2 Non-Biodegradation Processes

The few reports on bioremediation of colored effluents by yeasts usually mention nonenzymatic processes as the major mechanism for azo dye decolorization [5–10]. In a first approximation based on the cellular viability status, these processes can be divided into two different types: bioaccumulation and biosorption. Bioaccumulation usually refers to an active uptake mechanism carried out by living microorganisms (actively growing yeasts). The possibility of further dye biotransformation by redox reactions may also occur due to the involvement of the yeast metabolism. The main advantage of using bioaccumulating yeasts in color removal is avoiding the need for a separate biomass production step. However, we may anticipate possible limitations of employing live yeasts, such as potential toxic effects of the azo dyes and possible inability to deal with high strength effluents. On the other hand, the growth and performance of bioaccumulating yeasts will be mainly constrained by the nutrients' availability, notably carbon and nitrogen sources.

Biosorption is a general phenomenon that can occur in either dead or living biomass. However, this process usually refers to a passive uptake mechanism carried out by nonviable microorganisms (dead yeasts). The biosorption process involves physical–chemical interactions between the yeast surface and the azo dyes, as well as possible passive diffusion inside dead cells.

Using nonviable cellular biomass for azo dye removal has some advantages, namely the ability to function under extreme conditions of temperature and pH, and without addition of growth nutrients [10]. Also, waste yeast biomass, which is a by-product of industrial fermentations such as beer production, can be used as a relatively cheap source for biosorption of azo dyes. An important setback is the fact that the use of biomass for dye removal leads to an increase in the sludge amount, which requires further removal and treatment.

2.3 Biodegradation Processes

Since 1990 several reports have demonstrated the effectiveness of enzymatic decolorization and mineralization to carbon dioxide and water of azo dyes by fungi, notably white-rot basidiomycetous strains belonging to several genera [1, 11–14]. More recently, it was observed that a few ascomycetous yeast species such as *Candida zeylanoides* [15, 16], *Candida tropicalis*, *Debaryomyces polymorphus* [17], *Issatchenkia occidentalis* [18], *Saccharomyces cerevisiae* [19], *Candida oleophila* [20], and *Candida albicans* [21] perform a putative enzymatic biodegradation and concomitant decolorization of several azo dyes. The unique member of basidiomycetous yeasts allegedly performing a putative enzymatic biodegradation of azo dyes seems to be *Trichosporon* sp. (closely related to the *Trichosporon multisporum–Trichosporon laibachii* complex), which has been recently identified and characterized by [22, 23].

The yeast-mediated enzymatic biodegradation of azo dyes can be accomplished either by reductive reactions or by oxidative reactions. In general, reductive reactions led to cleavage of azo dyes into aromatic amines, which are further mineralized by yeasts. Enzymes putatively involved in this process are NADH-dependent reductases [24] and an azoreductase [16], which is dependent on the extracellular activity of a component of the plasma membrane redox system, identified as a ferric reductase [19]. Recently, significant increase in the activities of NADH-dependent reductase and azoreductase was observed in the cells of *Trichosporon beigelii* obtained at the end of the decolorization process [25].

The oxidative cleavage of azo dyes can be achieved by the action of the so-called ligninolytic enzymes laccase, manganese-dependent peroxidase, and lignin peroxidase. In general, the oxidation of azo dyes by ligninolytic enzymes led to the formation of a carbonium ion, and after nucleophilic water attack, a benzoquinone and a diazene-derivative are formed. Finally, the diazene is oxidized loosening molecular nitrogen to ultimately produce a hydroperoxide derivative [26]. Yang et al. [27] first found a good correlation between yeast-produced manganese-dependent peroxidase and the azo dye degradation (C.I. Reactive Black 5). Moreover, the presence of the azo dye in the culture medium was found to be indispensable for enzyme production by the yeast *D. polymorphus*. Recently, significant increase in the activities of lignin peroxidase (determined in assay reactions containing *n*-propanol as substrate) and other enzyme activities was observed in the cells of *S. cerevisiae* after decolorization of methyl red [24].

3 Diversity of Yeasts Involved in Azo Dyes Decolorization

To the best of our knowledge, the first work demonstrating the cleavage of the azo bond in a yeast-mediated process was published in the middle of last century by [28]. However, the practical interest in removing azo dyes with yeasts began several years later in the nineties [29, 30], where the biodegradation of several azo dyes was tested in *Candida curvata* and *Geotrichum candidum*. In spite of the fact that most investigations of microbial azo dye degradation utilize nonyeast microorganisms, a growing number of research groups have reported on several yeast species capable of decolorizing azo dyes (Table 1).

One of the first reports on yeast-mediated color removal by a putative process of biosorption of azo dyes by yeast (*Rhodotorula* sp.) biomass belongs to [31]. Yeast species such as *Kluveromyces marxianus* removed the diazo dye remazol black B [10], *Candida catenulata* and *Candida kefyr* removed more than 90% of amaranth by biosorption [6]. Biosorption uptake of the textile azo dyes remazol blue, reactive black, and reactive red by *S. cerevisiae* and *C. tropicalis* varied according to the selected dye, dye concentration, and exposure time [5, 7]. In a recent screening work carried out by [32], from the 44 yeast strains tested for their decolorization ability, 12 of them removed the dye Reactive Brilliant Red K-2BP by biosorption, among them the following were identified: *S. cerevisiae*, *Saccharomyces uvarum*, *Torulopsis candida*, and *Saccharomycopsis lipolytica*.

During the search and review of publications concerning yeasts with decolorizing capacity for azo dyes, we found that often researchers have given different names to the same yeast species. Some authors gave the name of the anamorph (which is the asexual or mitosporic form), some the name of its teleomorph (which is the sexual or meiosporic form), and others an obsolete name. For example, *T. candida* is an obsolete name for *Candida famata*, the anamorph phase of *Debaryomyces hansenii*.

Table 1 Yeast species and biodegradation ability of azo dyes

Yeast	Azo dye	[Dye]	Dye removal (%)	References
Candida curvata[a] (immobilized)	Several (10) e.g., Orange II	Not referred	65.0–75.0	[29]
Geotrichum candidum	Reactive Red 33 Acid Red 73 Acid Blue 324 Reactive Black 5	100–200 ppm	Not referred	[30]
Candida zeylanoides	Several azo dyes	10–50 ppm	20–90.0	[15]
C. zeylanoides	Methyl Orange Orange II	0.2 mM	85–100.0	[16]
Geotrichum sp. (solid media)	Acid Red 183	0.1–1.0 (g/L)	Not referred	[40]
Geotrichum sp.	Reactive Black 5 Reactive Red 158 Reactive Yellow 27	100 (mg/L)	100.0 (variable, time dependent)	[13]
Debaryomyces polymorphus	Reactive Black 5	100 (mg/L)	94.0	[17]
Candida tropicalis			97.0	
D. polymorphus	Reactive Red	100 (mg/L)	69.0	[17]
C. tropicalis			30.0	
D. polymorphus	Reactive Yellow	100 (mg/L)	70.0	[17]
C. tropicalis			40.0	
D. polymorphus	Reactive Brilliant Red	100 (mg/L)	85.0	[17]
C. tropicalis			40.0	
Issatchenkia occidentalis	Methyl Orange Orange II	0.2 (mmol/L)	>95.0 85.0	[18]
Saccharomyces italicus[b] *Saccharomyces chevalieri*[b] *Torulopsis candida*	Reactive Brilliant Red	50 (mg/L)	~87.0	[32]
Candida krusei *Pseudozyma rugulosa*	Reactive Brilliant Red	50–200 (mg/L)	99.9	[32]
C. krusei	Acid Brilliant Red B	50 (mg/L)	62.0–94.0	[32]
P. rugulosa	Reactive Black KN-B Acid mordant yellow		22.0–98.0	
D. polymorphus	Reactive Black 5	100–1,000 (mg/L)	95.0–98.0	[27]
Candida oleophila	Reactive Black 5	50–200 (mg/L)	95.0–100.0	[20]
Trichosporon multisporum	Reactive Red 141	200 (mg/L)	90.2–94.5	[22]
T. multisporum/T. laibachii complex			89.8–92.8	[22]
			100.0	[23]

(*continued*)

Table 1 (continued)

Yeast	Azo dye	[Dye]	Dye removal (%)	References
Geotrichum sp. (immobilized)	Orange G	100 (mg/L)	>96.0	[41]
Galactomyces geotrichum[c]	Methyl Red	100 (mg/L)	100.0	[42]
	Amido Black 10B	50 (mg/L)	92.0	
Yeast consortium	Reactive Violet 5	20–100 (mg/L)	78.2–89.3	[43]
	Reactive Orange 16		53.1–99.5	
Candida albicans	Direct Violet 51	100 (mg/L)	73.2	[21]
T. beigelii (syn. *T. cutaneum*)	Reactive Blue 171	50 (mg/L)	100.0	[25]
	Reactive Red 141		85.0	
	Reactive Green 19 A		70.0	
	Reactive Yellow 17		60.0	
	Reactive Orange 94		50.0	
Fungal consortium[d]	Reactive Black 5	30 (mg/L)	70.0–80.0	[44]
	Reactive Red	Dilution: 200–320×	65.0	
	Acid Red 249		94.0	
	Textile wastewater		89.0	

Unless otherwise stated, all experiments were done in liquid media and with free biomass
[a]Obsolete name for *Cryptococcus curvatus*
[b]Synomyms of *Saccharomyces cerevisiae*
[c]Obsolete synonyms: *Endomyces geotrichum* and *Dipodascus geotrichum*
[d](21 fungal strains), 70% of them belongs to *Candida* genus

From the literature consultation it is worth noting that the majority of the yeast species involved in azo dye decolorization belongs to the Ascomycota phylum. Additionally, it seems that the azo dyes color removal ability is restricted to Saccharomycetales order. By contrast, and in spite of fewer reports involving basidiomycetous yeasts in azo dyes degradation, three yeast species are scattered in two different orders (Table 2).

This Ascomycota dominance in scientific literature contradicts the results obtained by us in a very recent screening for the azo dye color removal abilities of 92 wild yeast isolates (77% Basidiomycota) recovered from decomposing leaves in a freshwater marsh. The 12 best isolates, tested in their abilities to remove the azo dyes C.I. Reactive Black 5, C.I. Reactive Violet 5, C.I. Acid Red 57, C.I. Reactive Orange 16, and Methyl Orange, exhibiting at least full decolorization of two of them, were identified by molecular methods. The identification showed that the isolates belong to seven species, six Basidiomycota: *Filobasidium* sp. (order Filobasidiales), *Rhodosporidium kratochvilovae*, *Rhodotorula graminis* (order Sporidiobolales), *Cryptococcus laurentii*, *Cryptococcus podzolicus*, and *Cryptococcus victoriae* (order Tremellales) and one Ascomycota, *Candida parapsilosis* (order Saccharomycetales).

In view of the present state of the art, we think that it is important to test other yeast species, either from Ascomycota or Basidiomycota phylum, to know the real diversity of yeasts capable of removing azo dyes, their main mechanisms of decolorization, and biotechnological potential.

Table 2 Taxonomy of azo dye-decolorizing yeast species and their anamorph/teleomorph correspondent names

Phylum	Order	Anamorph name	Teleomorph name
Ascomycota	Saccharomycetales	*Candida albicans*	Unknown
		Candida famata	*Debaryomyces hansenii*
		Candida krusei	*Issatchenkia orientalis*
		Candida oleophila	Unknown
		Candida robusta	*Saccharomyces cerevisiae*
		Candida sorbosa	*Issatchenkia occidentalis*
		Candida tropicalis	Unknown
		Candida zeylanoides	Unknown
		Geotrichum candidum[a]	*Galactomyces geotrichum*
		Geotrichum candidum (group A)	*Galactomyces candidus*
Basidiomycota	Ustilaginales	*Pseudozyma rugulosa*	Unknown
	Trichosporonales	*Trichosporon multisporum*[b]	Unknown
	Trichosporonales	*Cryptococcus curvatus*[c]	Unknown

[a]Sensu strictum. Other groups of *Galactomyces geotrichum/Geotrichum candidum* complex contain more three species (see [45])
[b]Fell et al. [46]
[c]Biswas et al. [47]

4 Combination of Chemical Pretreatment with Yeasts for Azo Dyes Decolorization

The biological treatment of wastewaters containing organic compounds, like azo dyes, is not an easy process due to the refractory character of some of them. The difficulties and even failures in the biological removal of azo dye compounds strongly suggest the use of a previous chemical pretreatment process [33], mainly because bioremediation usually do not achieve full degradation of recalcitrant compounds. Several factors can affect the biodegradation process. It may depend on the environmental conditions and on the nature of chemical compounds to be degraded. Two main factors have been identified responsible for the bio-recalcitrant behavior of some organic compounds: the lack of enzymes that are able to degrade the molecule (which will depend on the size, nature, number, and position of functional groups) and its toxic properties against live yeast cells (the capability to disrupt vital functions or even to produce the death of the microorganisms). In this sense, organic compounds may be non-biocompatible due to their toxic or non-biodegradable character.

Previous studies have attempted the strategy of combining chemical and biological processes to treat contaminants in wastewaters. These studies, extensively reviewed by [33], suggested potential advantages for the field of effluents treatment. Recently, some interesting coupled systems, advanced oxidation processes (AOPs)–biological agents (cells or enzymes), have been proposed to treat various types of industrial wastewaters. Textile, pulp and paper, surfactants, explosives from military industries, phenolic-rich agro-industrial effluents, and pesticides contaminated effluents are some examples of wastewaters treated with combined processes.

It can be said that the use of AOPs in conjunction with biological oxidation has been a recent innovation in the treatment strategies for wastewater. A major drawback of AOPs is their relatively high operational costs compared to those of biological treatments. However, the use of AOPs as a pretreatment step for the enhancement of biodegradability of wastewater containing recalcitrant or inhibitory compounds can be justified when the intermediates resulting from the reaction can be readily degraded by microorganisms. Therefore, combinations of AOPs as preliminary treatments with low-cost biological processes seem very promising from an economical point of view [33, 34].

To the best of our knowledge, the only work until now published that combines a chemical process with a biological process using yeasts was presented by [35]. This study describes the employment of an AOP – Fenton's reagent – as a pretreatment for further aerobic treatment with the yeast *C. oleophila* in the decolorization of the azo dye C.I. Reactive Black 5 (RB5).

The major purpose of this integrated process was to reduce the operational costs, particularly the hydrogen peroxide concentration used in Fenton's reagent, to decolorize a RB5 concentration of 500 mg/L. The study was conducted to evaluate the efficiency of Fenton's reagent as pretreatment during 60 min, performed at different hydrogen peroxide dosages. After that, each Fenton pretreated effluent was inoculated with viable cells of the yeast *C. oleophila* to remove the remaining concentration of RB5.

Combining Fenton's reagent and *C. oleophila* yeast, a total color removal of 91 and 95% was achieved for an initial hydrogen peroxide concentration of 1.0 and 2.0 mmol/L, respectively. Moreover, it should be pointed out that by doubling the initial hydrogen peroxide concentration, only a minor impact was obtained in the final dye decolorization. However, using Fenton's reagent alone, much higher hydrogen peroxide concentration (5.0 mmol/L) was necessary to achieve identical color removal.

Since optimal hydrogen peroxide concentration could be selected and according to operational costs, an effective RB5 decolorization process can be reached by combining an AOP (Fenton's reagent) and a yeast treatment (viable cells of *C. oleophila*) under aerobic conditions. According to similar works [36], Fenton's reagent is an efficient process to improve the biodegradability of organic pollutants. Therefore, to reduce costs, the main goal should not be to obtain a complete decolorization of the azo dye solution with the chemical process, but the generation of a more biodegradable effluent for further biological treatment.

5 Conclusion and Perspectives

Only a restrict group of microorganisms are able to bring about the complete biodegradation of recalcitrant polluting compounds, azo dyes being a case in point. It is interesting to point out that yeasts can also be involved in lignin (an aromatic and highly recalcitrant biopolymer) biodegradation such as *Candida* sp. [37], *Rhodotorula glutinis* [38], and *Trichosporon cutaneum* [39].

In recent years, a growing interest in the research devoted to the biodegradation of azo dyes have been putting in evidence both the feasibility of yeast-mediated decolorization and the metabolic versatility exhibited by yeasts. Taken together, these are very encouraging findings, since the majority of yeast species have never been screened for azo dyes bioremediation.

References

1. Fu Y, Viraraghavan T (2001) Fungal decolorization of dye wastewaters: a review. Biores Technol 79:251–262
2. Kirby N, Mc M, Marchant R (1995) Decolorization of an artificial textile effluent by *Phanerochaete chrysosporium*. Biotechnol Lett 17:761–764
3. Barnett JA, Payne RW, Yarrow D (2000) Yeasts: characteristics and identification, 3rd edn. Cambridge University Press, Cambridge, UK
4. Kurtzman CP, Fell JW (1998) The Yeasts: a taxonomic study, 3rd edn. Elsevier Science, New York
5. Aksu Z (2003) Reactive dye bioaccumulation by *Saccharomyces cerevisiae*. Process Biochem 38:1437–1444
6. Camargo AT, Corso CR (2002) Remoção do corante amaranto (C.I. 16.185) por biomassa de leveduras do género Candida através de biosorção. Biol Health Sci 8:75–85
7. Dönmez G (2002) Bioaccumulation of the reactive textile dyes by *Candida tropicalis* growing in molasses medium. Enzyme Microb Technol 30:363–366
8. El-Sharouny EE, El-Sersy NA (2005) Biosorption of textile dyes by brewer's yeast biomass. Fresenius Environ Bull 14:928–933
9. Ertuğrul S, San NO, Dönmez G (2009) Treatment of dye (Remazol Blue) and heavy metals using yeast cells with the purpose of managing polluted textile wastewaters. Ecol Eng 35:128–134
10. Meehan C, Banat IM, McMullan G et al (2000) Decolorization of remazol black-B using a thermotolerant yeast, *Kluyveromyces marxianus* IMB3. Environ Int 26:75–79
11. Cripps C, Bumpus JA, Aust SD (1990) Biodegradation of azo and heterocyclic dyes by *Phanerochaete chrysosporium*. Appl Environ Microbiol 56:1114–1118
12. Gill PK, Arora DS, Chander M (2002) Biodecolorization of azo and triphenylmethane dyes by *Dichomitus squalens* and *Phlebia* spp. J Ind Microbiol Biotechnol 28:201–203
13. Máximo C, Amorim MTP, Costa-Ferreira M (2003) Biotransformation of industrial reactive azo dyes by *Geotrichum* sp. CCMI 1019. Enzyme Microb Technol 32:145–151
14. Selvam K, Swaminathan K, Chae K-S (2003) Microbial decolorization of azo dyes and dye industry effluent by *Fomes lividus*. World J Microbiol Biotechnol 19:591–593
15. Martins MAM, Cardoso MH, Queiroz MJ (1999) Biodegradation of azo dyes by the yeast Candida zeylanoides in batch aerated cultures. Chemosphere 38:2455–2460
16. Ramalho PA, Scholze H, Cardoso MH et al (2002) Improved conditions for the aerobic reductive decolourisation of azo dyes by *Candida zeylanoides*. Enzyme Microb Technol 31:848–854
17. Yang Q, Yang M, Pritsch K et al (2003) Decolorization of synthetic dyes and production of manganese-dependent peroxidase by new fungal isolates. Biotechnol Lett 25:709–713
18. Ramalho PA, Cardoso MH, Cavaco-Paulo A et al (2004) Characterization of azo reduction activity in a novel ascomycete yeast strain. Appl Environ Microbiol 70:2279–2288
19. Ramalho PA, Paiva S, Cavaco-Paulo A et al (2005) Azo reductase activity of intact *Saccharomyces cerevisiae* cells is dependent on the Fre1p component of plasma membrane ferric reductase. Appl Environ Microbiol 71:3882–3888

20. Lucas MS, Amaral C, Sampaio A et al (2006) Biodegradation of the diazo dye Reactive Black 5 by a wild isolate of *Candida oleophila*. Enzyme Microb Technol 39:51–55
21. Vitor V, Corso CR (2008) Decolorization of textile dye by *Candida albicans* isolated from industrial effluents. J Ind Microbiol Biotechnol 35:1353–1357
22. Pajot HF, Figueroa LIC, Fariña JI (2007) Dye-decolorizing activity in isolated yeasts from the ecoregion of Las Yungas (Tucumán, Argentina). Enzyme Microb Technol 40:1503–1511
23. Pajot HF, Figueroa LIC, Spencer JFT et al (2008) Phenotypical and genetic characterization of *Trichosporon* sp. HP-2023 a yeast isolate from Las Yungas rainforest (Tucumán, Argentina) with azo-dye-decolorizing ability. Anton van Leeuw 94:233–244
24. Jadhav JP, Parshetti GK, Kalme SD et al (2007) Decolourization of azo dye methyl red by *Saccharomyces cerevisiae* MTCC-463. Chemosphere 68:394–400
25. Saratale RG, Saratale GD, Chang JS et al (2009) Decolorization and biodegradation of textile dye Navy blue HER by *Trichosporon beigelii* NCIM-3326. J Hazard Mater 166:1421–1428
26. Chivukula M, Renganathan V (1995) Phenolic azo dye oxidation by laccase from *Pyricularia oryzae*. Appl Environ Microbiol 61:4374–4377
27. Yang Q, Yediler A, Yang M et al (2005) Decolorization of an azo dye, Reactive black 5 and MnP production by yeast isolate: *Debaryomyces polymorphus*. Biochem Eng J 24:249–253
28. Mecke R, Schmähl D (1957) Die spaltbarkeit der azo-brücke durch hefe (Cleavage of azo bridge links by yeast). Arzneim-Forsch 7:335–340
29. Kakuta T, Tateno Y, Koizumi T et al (1992) Azo dye waste-water treatment with immobilized yeast. J Soc Ferment Technol 70:387–393
30. Kim SJ, Ishikawa K, Hirai M et al (1995) Characteristics of a newly isolated fungus, *Geotrichum candidum* Dec 1, which decolorizes various dyes. J Ferment Bioeng 79:601–607
31. Trindade RC, Angelis DF (1995) Removal of azo dyes for *Rhodotorula*: relationships with pH and substantivity index. 7th International Symposium on Microbial Ecology, Santos, S. Paulo, Brazil Abstract P3-24.86
32. Yu Z, Wen X (2005) Screening and identification of yeasts for decolorizing synthetic dyes in industrial wastewater. Int Biodeterior Biodegrad 56:109–114
33. Scott J, Ollis D (1995) Integration of chemical and biological oxidation processes for water treatment: review and recommendations. Environ Progr 14:88–103
34. Litter MI (2005) Introduction to photochemical advanced oxidation processes for water treatment. In: Boule P, Bahnemann DW, Robertson PKJ (eds) Environmental Photochemistry Part II, vol 2. Springer, Berlin/Heidelberg, pp 325–366
35. Lucas MS, Dias AA, Sampaio A et al (2007) Degradation of a textile reactive azo dye by a combined chemical-biological process: Fenton's reagent-yeast. Water Res 41:1103–1109
36. Chamarro E, Marco A, Esplugas S (2001) Use of Fenton reagent to improve organic chemical biodegradability. Water Res 35:1047–1051
37. Clayton NE, Srinivasan VR (1981) Biodegradation of Lignin by *Candida* spp. Naturwissenschaften 68:97–98
38. Gupta JK, Sharma P, Kern HW et al (1990) Degradation of synthetic lignins and some lignin monomers by the yeast *Rhodotorula glutinis*. World J Microb Biotechnol 6:53–58
39. Georgieva N, Yotova L, Betcheva R et al (2006) Biobleaching of lignin in linen by degradation with *Trichosporon cutaneum* R57. J Univ Chem Technol Metall 41:153–156
40. Jarosz-Wiłkoazka A, Kochmaska-Rdest J, Malarcyk E et al (2002) Fungi and their ability to decolourize azo and anthraquinonic dyes. Enzyme Microb Technol 30:566–572
41. Zeroual Y, Kim BS, Yang MW et al (2007) Decolorization of some azo dyes by immobilized *Geotrichum* sp. biomass in fluidized bed bioreactor. Appl Biochem Biotechnol 142:307–316
42. Jadhav SU, Kalme SD, Govindwar SP (2008) Biodegradation of methyl red by *Galactomyces geotrichum* MTCC 1360. Int Biodeterior Biodegrad 62:135–142
43. Olteanu Z, Roşu CM, Mihăşan M et al (2008) Preliminary consideration upon oxide-reductive system involved in aerobic biodegradation of some textile dyes. Analele Ştiinţifice ale Universiţătii Alexandru Ion Cuza, Secţiunea Genetică şi Biologie Moleculară, TOM IX, 41–46

44. Yang Q, Lib C, Li H et al (2009) Degradation of synthetic reactive azo dyes and treatment of textile wastewater by a fungi consortium reactor. Biochem Eng J 43:225–230
45. De Hoog G, Smith M (2004) Ribosomal gene phylogeny and species delimitation in *Geotrichum* and its teleomorphs. Stud Mycol 50:489–515
46. Fell JW, Boekhout T, Fonseca A et al (2000) Biodiversity and systematics of basidiomycetous yeasts as determined by large-subunit rDNA D1/D2 domain sequence analysis. Int J Syst Evol Microbiol 50:1351–1371
47. Biswas SK, Wang L, Yokoyama K et al (2005) Molecular phylogenetics of the genus *Trichosporon* inferred from mitochondrial Cytochrome b gene sequences. J Clin Microbiol 43:5171–5178

Factors Affecting the Complete Mineralization of Azo Dyes

Laura Bardi and Mario Marzona

Abstract Azo dyes are complex compounds generally recalcitrant to biodegradation. From their catabolism several toxic and carcinogenic compounds are formed, in particular when their decolorization is reached through a reductive cleavage of the azo groups. For this reason the full degradation of the dyes and the intermediates is necessary to prevent risks for human health. Their mineralization can usually be reached with aerobic treatments or with two-steps anaerobic/aerobic treatments. Several environmental and physiological factors can influence the microbial activity and consequently the efficacy and effectiveness of the complete biodegradation processes. The roles of oxygen, bioavailability, adsorption, nutrients and cometabolic induction, dye concentration, pH, temperature, and salinity are treated.

Keywords Azo dyes, Bioavailability, Biodegradation, Cometabolism, Environment, Oxygen

Contents

L. Bardi (✉)
C.R.A. – Agricultural Research Council, R.P.S. Research Centre for the Soil-Plant System, R.G. Turin Via Livorno 60 (A2 laboratori), 10144 Turin, Italy
e-mail: laura.bardi@entecra.it

M. Marzona
Department of General and Applied Organic Chemistry, University of Turin, Corso Massimo D'Azeglio, 48, 10125 Turin, Italy

H. Atacag Erkurt (ed.), *Biodegradation of Azo Dyes*,
Hdb Env Chem (2010) 9: 195–210, DOI 10.1007/698_2009_50,

Abbreviations

FMN	Flavin mononucleotide
LiP	Lignin peroxidases
MnP	Manganese peroxidases
NADH	Nicotinamide adenine dinucleotide
NADPH	Nicotinamide adenine dinucleotide phosphate
TNT	Trinitrotoluene
VP	Versatile peroxidase

1 Introduction

Azo dyes are the largest class of synthetic dyes. Among the colorants indexed in the Color Index, published by the Society of Dyers and Colorists, about 70% are azo dyes, followed by anthraquinones (about 15%). They are largely used to color textiles, cosmetics, leather, pharmaceuticals, paints, printing inks, plastics, and foods, and they are also used as biological stains in laboratories and clinics [1].

The amount of dyestuff that does not bind to the fibers and is lost in wastewater during textile processing is estimated from 5–10% to 50% in the case of reactive dyes [2], and the azo dye concentration in wastewater produced by textile industries varies from 5 to 1,500 mg L^{-1} [3]. Azo dyes are xenobiotics strongly recalcitrant to biological degradation processes: they are not degraded in conventional aerobic sewage treatment plants and became a great environmental hazard [4]. Azo dyes released in the environment are an important risk for human health, as a potential source of carcinogenic aromatic amines. Azo dyes can enter the human body through the food chain or by skin contact; in the liver and in the gastrointestinal tract, they are reduced by azoreductases to aromatic amines, which induce urinary bladder cancer in humans and tumors in experimental animals. The mechanisms of carcinogenic activation of azo dyes are reduction and cleavage of the azo bond, oxidation of azo dyes with structures containing free aromatic amines groups and direct oxidation of the azo linkage to highly reactive electrophilic diazonium salts [1]. Therefore, to avoid the risk for human health due to azo dyes, their complete degradation is of main importance: decolorization is not sufficient, if their metabolites are not completely mineralized. To reach the complete biodegradation, several environmental and physiological factors acting on microbial metabolisms have to be taken into account.

2 Classification of Azo Dyes

All azo dyes contain one or more azo groups (–N=N–) as chromophore in the molecule; on the basis of the number of azo groups in each molecule, they are named monoazo-, disazo-, trisazo-, etc. The azo groups are in general bound to a benzene or naphthalene ring, but they can also be attached to heterocyclic aromatic molecules or to enolizable aliphatic groups. On the basis of the characteristics of the processes in which they are applied, the molecule of the dye is modified to reach the best performances; so they can be acid dyes, direct dyes, reactive dyes, disperse dyes, or others.

2.1 Acid Dyes

Acid dyes constitute a large group of water-soluble anionic colorants with relatively low molecular weights, typically characterized by the presence of strongly water-solubilizing substituents, especially sulfonate groups. They are mainly composed of aromatic monoazo compounds, but they also include bisazo, nitro, 1-aminoanthraquinone triphenylmethine, and other groups of dyes. Aromatic sulfonates are not only easily accessible synthetically, but also have the advantage of being negatively charged in aqueous solution over an extremely broad pH range. Anionic monoazo dyes and their metal salts are widely used for either dyeing paper and leather or as pigments. Their main application, however, constitutes the dyeing of proteins, that is animal hair fibers (wool, silk) and synthetic fibers (nylon). In this context, the term acid dyes is often used, since the corresponding dyeing process takes place in a weakly acidic solution (pH 2–6). Attachment to the fiber is attributed, at least partly, to the salt formation between anionic groups in the dyes and cationic groups in the fiber: animal protein fibers and nylon fibers contain many cationic sites. A certain amount of dyestuff always remains in water after dyeing.

2.2 Direct Dyes

Direct dyes are attracted to the textile, according to their "substantivity," by intermolecular forces without the need of mordant. They are used to color cotton and paper leather, silk, and nylon, and are also used as pH indicators or as biological stains. The water solubility is assured by sulfonate groups (usually 2–4), and direct dyeing is normally carried out in a neutral or slightly alkaline dyebath; washing is easy and fast.

2.3 Reactive Dyes

Reactive dyes contain substituent that, when activated, react with the –OH groups of cellulose (i.e., cotton) or with $-NH_2$ and –SH groups of protein fibers (i.e., wool) forming covalent bonds, making them among the most permanent of dyes.

2.4 Disperse Dyes

Disperse dyes are almost insoluble in water; they do not contain any basic or acidic group in the molecule. They are finely ground mixed to a dispersing agent and disposed as powder or paste, and then used as aqueous suspensions. They are usually used to dye cellulose acetate, nylon, triacetate and polyester fibers; also acrylics can be dyed with disperse dyes, but with poor intensity. High temperature and pressure of dyebath required is in some cases, and dyeing rate is influenced by the particle size and the chosen dispersing agent.

3 Factors Affecting the Complete Mineralization of Azo Dyes

3.1 Oxygen Availability

As extensively treated in other chapters, the most common way to reach the complete azo dye mineralization consists of two steps: a first step in which a reductive cleavage of the diazo bond gives rise to the production of colorless metabolites, mainly aromatic amines, and a second step in which the resulting metabolites are degraded in aerobiosis. The first step usually occurs in anaerobic conditions, but it can also be carried out by several aerobic bacteria, which are able to synthesize azoreductases cleaving the azo group in the presence of molecular oxygen. Under aerobic conditions, fungal degradation of azo dyes has also been described [4]. Moreover, the complete degradation of azo dyes without the release of aromatic amines has also been observed in a single, anoxic step with anaerobic bacteria such as *Clostridium bifermentans*; as *Clostridium* spp. are able to degrade aromatic compounds, such as cyclic nitroamines and TNT, it was proposed that the metabolic byproducts formed by this microorganism are different from the aromatic amines produced by other anaerobic bacteria [5].

The extent to which azo dyes are reduced is dependent on the electron density around the diazo bond: when the electron density is decreased, the azo group is more easily reduced and an aromatic amine is released. Electron density is decreased by the electron-withdrawing groups, while a charged functional group in the proximity of the azo group or the presence of a second polar group interferes with the reaction. Electron-donor groups in ortho-position with respect to diazo groups cause a reduction through the formation of hydrogen bonds. A simpler reduction of the diazo group is also observed in water-soluble dyes that contain groups such as $-SO_3Na$ or –COOH [1].

Redox mediators, such as flavins or quinones, are usually involved in the azo bond reduction. Therefore, the azo bond cleavage is a chemical, unspecific reaction that can occur inside or outside the cell, relying on the redox potential of the redox mediators and of the azo compounds. Also the reduction of the redox mediators can be both a chemical and an enzymatic process. As a consequence, it is an evidence that environmental conditions can affect the azo dyes degradation process extent both directly, depending on the reductive or oxidative status of the environment, and indirectly, influencing the microbial metabolism.

Anaerobiosis is the most studied environmental factor affecting this reaction. Under aerobic conditions, oxygen, and azo dyes are in competition for the reduced electron carriers.

From the biological point of view, the effect of anaerobiosis has been characterized in purely anaerobic, facultative anaerobic, and aerobic bacteria, in yeasts, and in tissues from higher organisms [6–12]. From these studies it can be deduced that almost every azo compound can be biologically reduced under anaerobic conditions [4]. Reduced flavins are produced by cytosol flavin-dependent reductases [6, 13], while quinone reductase activity located in the plasma membrane [14] and extracellular azo reductase activities [9, 15] were also observed.

Bacterial aerobic azoreductases have also been described, belonging to four different families [1]. The four enzyme families can be divided into two groups: flavin-free enzymes, using NADPH [16, 17] or NADH [18] as cofactors, belong to one group; flavin-dependent azoreductases belong to the other group [19–23]. A FMN-reductase with minor activity on Ethyl Red cleavage and a plasma membrane ferric reductase activity were also described in *Saccharomyces cerevisiae* [24, 25].

The degradation of azo dyes in aerobiosis can also be carried out by lignin-degrading fungi, mainly white-rot fungi, or by peroxidase-producing bacterial strain, mainly *Streptomyces* species, as extensively reviewed by Stolz [4].

3.2 Bioavailability

The different chemical–physical characteristics of the dyes molecules, as well as of their intermediates, can differently influence the bioavailability through the actual concentration in the aqueous phase, where microorganisms or enzymes are active, or through their potentiality to pass through the plasma membrane to be metabolized inside the cell.

The hydrophylicity or hydrophobicity are main factors influencing the fate of azo dyes when they come in contact with living organisms. The water-soluble azo compounds, such as sulfonated azo dyes, are highly polar molecules that cannot pass the plasma membrane barrier. Indeed only the biological systems in which the enzymes of the catabolic pathway, or the redox mediators responsible of the reductive cleavage of the azo bond, are extracellular are effective for the degradation of these compounds. The hydrophobic azo compounds that are fat-soluble, such as Sudan azo dyes, can pass the plasma membrane barrier and can be degraded in the cytoplasm; they are easily adsorbed through the skin, but their availability in the aqueous phase, in which the degradative microorganisms are active, is low.

The bioavailability can be improved by compounds that increase the water solubility, but very few assays have been carried out in this direction. Liposomes are effective inducing a faster decolorization of Acid Orange 7 by anaerobic biomass [26]. A strategy of selection of microbial strains able to improve the bioavailability of insoluble dyes can also be carried out: a *Shewanella* strain J18 143 was characterized for its ability to degrade large pigment aggregates of dispersed dye to produce individual pigment particles [27].

To reach the reductive step of the azo bond cleavage, due to the reaction between reduced electron carriers (flavins or hydroquinones) and azo dyes, either the reduced electron carrier or the azo compound should pass the cell plasma membrane barrier. Highly polar azo dyes, such as sulfonated compounds, cannot pass the plasma membrane barrier, as sulfonic acid substitution of the azo dye structure apparently blocks effective dye permeation [28]. The removal of the block to the dye permeation by treatment with toluene of *Bacillus cereus* cells induced a significant increase of the uptake of sulfonated azo dyes and of their reduction rate [29]. Moreover, cell extracts usually show to be more active in anaerobic reduction of azo dyes than whole cells. Therefore, intracellular reductases activities are not the best way to reach sulfonated azo dyes reduction; the biological systems in which the transport of redox mediators or of azo dye through the plasma membrane is not required are preferable to achieve their degradation [13].

The anaerobic reduction of azo dyes mediated by quinones was reached with cell culture supernatants of *Sphingomonas xenophaga* BN6 [14]. The addition of quinones (anthraquinone-2-sulfonate, 2-hydroxy-1,4-naphthoquinone) induced a significant increase of dye decolorization, and quinones acted as redox mediators and were reduced by reductase activity located in the cell membranes. The anaerobic reduction of azo dyes due to enzymatic activities bound to cell membranes was also observed in mammalian cells; NAD(P)H-cytochrome c reductase or cytochrome P450 system were involved [30, 31]. Another model useful for the reduction of sulfonated azo dyes, which does not require the transport through the plasma membrane, was observed in intestinal strictly anaerobic bacteria, producing extracellular azoreductases [9, 15]. In a work carried out with *Clostridium perfringens*, the extracellular reductive activity resulted independent from added flavins [9]. Also reduced inorganic compounds, produced as end products of anaerobic microbial metabolisms, can act as electron carriers involved in the reductive azo bond cleavage [4]. With sulfate-reducing bacteria, the decolorization of an azo dye (Reactive Orange 96) was observed associated with the formation of H_2S [16].

The reductive cleavage of sulfonated azo dyes by several bacterial strains in aerobiosis and in the presence of other carbon and energy sources has been observed [22, 32–36].

Extracellular peroxidases are produced by *Streptomyces chromofuscus*, with the capability to decolorize azo dyes associated to ligninolytic activity in aerobiosis. Azo dyes are converted to cationic radicals, which are subjected to nucleophilic attack by water or hydrogen peroxide molecules, producing reactive compounds that undergo redox reactions that result in a more stable intermediate [37].

Several fungi (i.e., *Phanerochaete chrysosporium*, *Geotricum candidum*, *Trametes versicolor*, *Bjercandera adusta*, *Penicillium* sp., *Pleurotus ostreatus*, *Pycnoporus cynnabarinus*, *Pyricularia oryzae*) are able to degrade azo dyes by extracellular enzymes production: lignin-degrading fungi are able to decolorize several kinds of even complex molecules of sulfonated and nonsulfonated azo dyes, mainly by synthesizing esoenzymes such as lignin peroxidases (LiP), manganese peroxidases (MnP) and laccases [1, 4]. In *Pleurotus sajor-caju* cultures grown over solid medium, it was observed that soluble azo dyes were degraded, even if incompletely, while insoluble azo dyes were not degraded during mycelial growth [38].

Also ascomycetes yeast strains showed decolorizing behaviors due to extracellular reactions on polar dyes. The process occur when an alternative carbon and energy source is available. The involvement of an externally directed plasma membrane redox system was suggested: in *S. cerevisiae*, the plasma membrane ferric reductase system participates in the extracellular reduction of azo dyes [25].

3.3 Adsorption

In conventional aerobic sewage-treatment plants, most azo dyes are not degraded, but about 40–80% of the dye physically adsorb to the sewage sludge [4].

Adsorption of azo dyes by the biomass is considered as the first step of their biological reduction [39]. Because of adsorption, the dye is concentrated onto the biomass until its saturation; the amount of adsorbed dye is then proportional to the amount of biomass [40–42]. Steffan et al. [43] observed that 68% Ethyl Orange was rapidly adsorbed on a microbial consortium immobilized in alginate beads, but only after the addition of glucose or starch the dye was effectively degraded.

Bacterial cells of *Oenococcus oeni* incubated for 48 h with three azo dyes (Fast red, Fast orange, and Methanil yellow) gave rise to decolorization due to adsorption, from 68% with Fast red to 30% with Fast orange and Methanil yellow [41]. Ozdemir et al. [44] observed a 93.9% decolorization of Acid Black 210 within 24 h by *Vibrio harveyi* TEMS1, a bioluminescent bacterium isolated from coastal seawater in Turkey. After extraction in methanol of biomass, the major part of the decolorized dye was recovered, indicating that decolorization was mainly due to adsorption. This assumption was confirmed by the fact that azoreductase activity was not detected. Khalid et al. [45] observed an adsorption on bacterial cells of Acid Red 88 and Disperse Orange 3, but not of Reactive Black 5 and Direct Red 81.

Initial adsorption of the dyes on fungal biomass followed by degradation was observed in cultures of *Irpex lacteus*, *Phanerochaete chrysosporium*, *Trametes versicolor*, and *Trichophyton rubrum* [46–49]. In *P. sajor-caju*, it was observed that Disperse Blue 79 and Acid Red 315 were incompletely or not degraded, but a decolorization was reached due to adsorption to the mycelium [38]. Also algae can be used as biosorbents of azo dyes [50].

3.4 Dye Concentration

An inverse relation between the efficiency of decolorization and the dye concentration has frequently been observed. This fact can be ascribed to several factors, the main of which can be considered the toxicity of the dyes at higher concentrations [41, 45, 51–53]. With Reactive Red 3B-A, concentrations from 100 to 2,000 ppm were tested with *C. bifermentans* [5]. At concentrations less than 200 ppm, 90% decolorization within 12 h was observed, while at very high dye concentration (>1,000 ppm), the decolorization rate decreased. Khalid et al. [54] observed an inverse relationship between the velocity of the decolorization reaction and the dye concentrations between 100 and 500 mg L^{-1} azo dye (Reactive Black 5, Direct Red 81, Acid Red 88, and Disperse Orange 3) by *Shewanella putrefaciens*. A decrease in decolorization percentage at a Acid Black 210 initial concentration growing from 100 to 400 ppm was also observed with *V. harveyi*, but the decrease was low [44].

3.5 Nutrients Availability and Cometabolic Induction

Many of the bacteria that decolorize azo dyes by reductive mechanisms need the availability of a supplementary carbon source, and therefore presumably do not use the azo dyes as the sole energy and carbon source [4]. Glucose is the most easily available and effective carbon source for microbial metabolism, and it has been frequently demonstrated that its addition improve the efficacy of azo dye degradation. Under anoxic conditions, carbon sources such as glucose, starch, fatty acids, yeast extract, tapioca, acetate, propionate, butyrate act as electron donors to allow the reduction of the azo bond [2, 55]. Ethanol was also successfully applied as electron donor for azo dye reduction during aerobic post-treatment: it created anaerobic microniches to facilitate anaerobic reduction in presence of oxygen [2]. Ozdemir et al. [44] tested the addition of glucose, fructose, maltose, sucrose, starch, and sodium acetate, in addition to yeast extract, to the growth medium of *V. harveyi* incubated under static conditions; the highest decolorization of Acid Black 210 was reached with glucose (92.1%) and the lowest with fructose (55.5%). The complete degradation of Reactive Red 3B-A, Reactive Black 5, and Reactive Yellow 3G–P was reached with *C. bifermentans* after the addition of glucose [5]. Also with *Oenococcus oeni* was observed an increase of decolorization percentage of Fast red from 66 to 93% due to 5 g L^{-1} glucose [41]. On the contrary, an inhibitory effect of glucose on azo dyes degradation was observed with a *S. putrefaciens* strain in high salinity conditions, which was ascribed to a preferential use of glucose for cell growth [54]. Xu et al. [12] found that lactate was the optimal carbon source among lactate, formate, glucose, and sucrose to remove Fast Acid Red GR with *Shewanella decolorationis* S12 under microaerophilic conditions; negligible dye decolorization was observed with resting cells without carbon sources. Yang et al. [56] observed a higher degradation of Reactive Black 5 by yeasts *Debaryomyces polymorphus* and *Candida tropicalis* corresponding to a higher concentration of glucose. The contemporary addition of sulfide as reductant with glucose produced an increase of Reactive Black 5 color removal [57].

The ability to use azo dyes as sole energy and carbon source by bacteria to be able to reduce the azo bond aerobically by a cometabolic way has been reported [2, 4]. A mixture of four structurally different dyes (Acid Red 88, Reactive Black 5, Direct Red 81, and Disperse Orange 3) was used as sole source of carbon and nitrogen to select six strains of bacteria tested for the ability to decolorize the dyes individually or in mixtures; a *S. putrefaciens* strain was identified as the most efficient [45].

Adaptation or cometabolic induction processes allow the degradation of azo dyes by several bacteria strains. The potential ability of bacteria to acquire novel metabolic traits was demonstrated adapting a bacterial consortium, which degraded 4,4′- dicarboxyazobenzene to the degradation of more complex azo compounds, such as carboxy-Orange I or carboxy-Orange II [58, 59]; however, the adaptation to the structurally analogous sulfonated dyes Acid Orange 20 and Acid Orange 7 was not possible [60]. A mutant strain of *Hydrogenophaga palleronii* able to grow on the sulfonate azo compound 4-carboxy-4′-sulfoazobenzene as the sole carbon and energy source until its mineralization was obtained by adaptation [61, 62]. In *P. sajor-caju* grown over solid medium, it was observed that the addition of dyes to culture medium induced a higher laccase activity, even if degradation did not occur; however, the addition of an antrachinoid dye (Reactive Blue 220) induced the degradation of azo dyes that otherwise were not degraded [38].

A stimulatory effect on cell growth due to nitrogen availability is usually related to a better azo dye degradation [63, 64]. However, an inhibitory effect of NH_4NO_3 on the decolorization of Reactive Black 5, Direct Red 81, Acid Red 88, and Disperse Orange 3 by *S. putrefaciens* was observed, and it was ascribed to the use of NO_3^- as a competing electron acceptor [54]. Several nitrogen sources (peptone, soya peptone, ammonium sulfate, ammonium chloride) were tested for Acid Black 210 decolorization with *V. harveyi*, showing a lower effect; the best performance was reached in LBM, composed by yeast extract, tryptone, and NaCl [44]. Similar results were found with *Pseudomonas luteola* on Reactive Red 22 [65]. Yeast extract is a complex organic substrate that provides carbon, nitrogen, and growth factors, and can be used by microbial cells as a source electron donor for reductive cleavage of the azo dyes [54]; it was found to be the best nitrogen source when tested for the decolorization of azo dyes [66]. The best decolorization of Everzold Red RBN by a bacterial consortium was obtained with yeast extract and lactose as nitrogen and carbon sources [51]. The maximum decolorization of Direct Red 81 was observed, with starch and casein as carbon and nitrogen sources [67].

Ligninolytic enzymes of the white rot fungi are usually expressed during secondary metabolisms, when carbon and nitrogen sources become limiting [4] and their expression is usually inducible. *Irpex lacteus*, which has been proven to be efficient in the degradation of synthetic dyes and other aromatic pollutants, has been studied to ascertain the role of manganese and of three synthetic dyes on the expression of different manganese-peroxidase isoenzymes secreted in crude culture liquids from mycelium immobilized in polyurethane foam [68]. It was observed that the specific degradation activity on different dyes changed in relation to the manganese concentration and to the dye molecule. Each manganese peroxidase isoform production and decolorization activity is differently regulated by different dyes, and their composition changed upon the aging of the cultures and depending on the structure of the dye added to the growth medium. The manganese peroxidases production was increased by high manganese concentration and by the addition of Bromophenol Blue, a triphenylmethane dye, but not by the Reactive Orange 16, an azo dye. The higher manganese level induced the production of new isoforms; moreover, the addition of dyes induced an effect in the isoform composition only at high manganese concentrations, while at low manganese concentration, no changes in the isoenzyme pattern were observed. However, the changes in isoenzyme profiles were observed after several days of cultivation, while 79% of Reactive Orange 16 was decolorized or adsorbed within 1 day; so it can be supposed that the enzyme production could be induced by metabolites produced from the dye degradation.

In *Phanerochaete flavido-alba*, an induction of ligninolytic activities that was ascribed to phenolic compounds was evidenced [69]. Phenols have also been shown to have an important role as redox mediators for dye degradation with laccases from *Pycnoporus cinnabarinus* and *Trametes villosa*, and they resulted to be necessary to degrade a strongly recalcitrant azo dye, the Reactive Black 5 [70].

Also in *Phanerochaete chrysosporium*, the effect of environmental factors on the expression of manganese-peroxidases has been well characterized [71]. At least three isoenzymes are active in the extracellular culture medium and the expression of each isomer is differently regulated by environmental factors: *mnp1* and *mnp2* genes are regulated by manganese in nitrogen-limited cultures, while *mnp3* gene seems not to be regulated by manganese availability; *mnp1* transcript predominates in agitated cultures, while *mnp2* transcript predominates in static cultures. These data support the hypothesis that the manganese peroxidase activity is assured under different culture conditions. Other factors affecting the expression of manganese peroxidases in *Phanerochaete chrysosporium* are the oxygen tension and the cell immobilization [72]. Immobilization in polyurethane foam has also been studied in *Nematoloma frowardii*, an agaric basidiomycete, in shaken flasks and aerated fermenter cultures with low nitrogen availability; the enzyme productivity in the immobilized cells was 1.4 times higher than that obtained with the free fungus, and the enzyme activity was not lost in recycling of the immobilized mycelium during three subsequent 10 day batches [73]. The production of ligninolytic enzymes by white rot fungi and the factors affecting their expression levels have been reviewed by Martinez [74]. It was reported that *Pleurotus* and *Bjerkandera* produce versatile peroxidases (VP), which are able to oxidize Mn^{2+} as well as nonphenolic aromatic compounds, phenols and dyes. Two VP genes have been cloned from *Pleurotus eryngii*, and it was reported that *Phanerochaete chrysosporium* manganese peroxidases and *P. eryngii* VP are induced by H_2O_2.

3.6 pH and Temperature

Several studies have been carried out to investigate the effect of pH on azo dye decolorization. In these assays, the decrease of absorbance at the wavelength corresponding to the maximum absorption for each dye is used as the method to evaluate the effectiveness of decolorization. Unfortunately, in most cases it is not clear if the isosbestic point of each dye was taken into account, and so it cannot be well understood if the different decolorization rate at different pH is due to a physical factor or to a differently influenced metabolic activity.

The tolerance to high pHs is important in particular for industrial processes using reactive azo dyes, which are usually performed under alkaline conditions. A strain of *C. bifermentans* selected from a contaminated soil was tested for the ability to decolorize Reactive Red 3B-A at pHs from 5 to 12; while no decolorization was observed at pH 5, the dye was nearly completely decolorized across a broad range of pH values (6–12) after 48 h of incubation; in this study a previous analysis of UV/Vis spectra of Reactive Red 3B-A, Reactive Black 5, and Reactive Yellow 3G-P after 0, 12, 24, and 36 h incubation was carried out, showing different decolorization rates for the three dyes, with no change in color content in the abiotic control [5].

Lignin peroxidases from ligninolytic fungi exhibit a 4.5–5 optimum pH [4]. In vitro assays with enzymatic extracts from *P. sajor-caju* grown in solid-state fermentation with sawdust of *Pinus* sp. and wheat bran were carried out to evaluate the effect of pH and temperature over the decolorization of Disperse Blue 79 and Acid Red 315. No significant differences were observed in decolorization due to pH; only for Disperse Blue 79 a greater decolorization at pH 5 and at 50°C was observed, with respect to pH 4.1, 6.0, and 6.5 at temperatures 30 and 40°C. In submerged cultures of the same fungus, incubated in presence of each dye under reciprocal agitation and under dark, a decrease of absorbance was observed, which was related to the decrease of pH of the growth medium due to metabolic activity.

The maximum rate of color removal is generally related to the optimum cell culture growth temperature for each microbial species, with an increase of decolorization proportional to the increase of temperature within the optimum temperature range [41, 42, 44, 75–78].

3.7 High Salinity

High salt concentrations up to 15–20% can be found in wastewater from dyestuff industries. Moreover, textile manufacturers located on coastal areas can cause pollution of seawater. The biological treatability of wastewater with a high saline concentration is limited because most of the microorganisms that are able to degrade azo dyes are not active in these conditions, in which the selection of halophilic or halotolerant bacteria capable to degrade azo dyes is necessary [79].

A highly salt-tolerant bacterial strain *Gracilibacillus* sp. GTY was tested for the ability to decolorize the azo dye Acid Red B. It was observed that the dye was decolorized by growing and resting cells, as well as by extracted azo reductase, in optimum conditions and at a 10–15% NaCl concentration; at very high and very low salt concentration, it was not possible to reach a good performance in decolorization [80].

Two bacterial *Shewanella* species, *S. putrefaciens* and *S. oneidensis*, previously selected on the basis of their ability to degrade azo dyes, were also tested in saline medium at different salt concentrations of up to 10% to evaluate their potential to decolorize four structurally different azo dyes: Reactive Black 5, Direct Red 81, Acid Red 88, and Disperse Orange 3. Full decolorization was reached at salt concentrations up to 6%; the decolorization velocity was inversely related to salt concentration. The rate of decolorization was increased by yeast extract and a calcium source, while was decreased by glucose and by a nitrogen source [54].

References

1. Chen H (2006) Recent advances in azo dye degrading enzyme research. Curr Protein Pept Sci 7(2):101–111
2. Rai HS, Bhattacharyya MS, Singh J, Bansal TK, Vats P, Banerjee UC (2005) Removal of dyes from the effluent of textile and dyestuff manufacturing industry: a review of emerging techniques with reference to biological treatment. Crit Rev Environ Sci Technol 35:219–238
3. Gottlieb A, Shaw C, Smith A, Wheatley A, Forsythe S (2003) The toxicity of textile reactive azo dyes after hydrolysis and decolorisation. J Biotechnol 101:49
4. Stolz A (2001) Basic and applied aspects in the microbial degradation of azo dyes. Appl Microbiol Biotechnol 56:69–80
5. Joe MH, Lim SY, Kim DH, Lee IS (2008) Decolorization of reactive dyes by *Clostridium bifermentans* SL186 isolated from contaminated soil. World J Microbiol Biotechnol 24:2221–2226
6. Adamson RH, Dixon RL, Francis FL, Rall DP (1965) Comparative biochemistry of drug metabolism by azo and nitro reductase. Proc Natl Acad Sci USA 54:1386–1391
7. Bragger JL, Lloyd AW, Soozandehfar SH, Bloomfield SF, Marriott C, Martin GP (1997) Investigations into the azo reducing activity of a common colonic microorganisms. Int J Pharmaceut 157:61–71
8. Dubin P, Wright KL (1975) Reduction of azo food dyes in cultures of *Proteus vulgaris*. Xenobiotica 5:563–571
9. Rafii F, Franklin W, Cerniglia CE (1990) Azoreductase activity of anaerobic bacteria isolated from human intestinal microflora. Appl Environ Microbiol 56:2146–2151

10. Scheline RR, Nygaard RT, Longberg B (1970) Enzymatic reduction of azo dye, Acid Yellow, by extracts of *Streptococcus faecalis*, isolated from rat intestine. Food Cosmet Toxicol 8:55–58
11. Wuhrmann K, Mechsner K, Kappeler T (1980) Investigation on rate-determining factors in the microbial reduction of azo dyes. Eur J Appl Microbiol 9:325–338
12. Xu M, Guo J, Sun G (2007) Biodegradation of textile azo dye by *Shewanella decolorationis* S12 under microaerophilic conditions. Appl Microbiol Biotechnol 76:719–726
13. Russ R, Rau J, Stolz A (2000) The function of cytoplasmic flavin reductases in the bacterial reduction of azo dyes. Appl Environ Microbiol 66:1429–1434
14. Kudlich M, Keck A, Klein J, Stolz A (1997) Localization of the enzyme system involved in the anaerobic degradation of azo dyes by *Sphyngomonas* sp. BN6 and effect of artificial redox mediators on the rate of azo reduction. Appl Environ Microbiol 63:3691–3694
15. Rafii F, Moore JD, Ruseler-van Embden JGH, Cerniglia CE (1995) Bacterial reduction of azo dyes used in foods, drugs and cosmetics. Microecol Ther 25:147–156
16. Zimmermann T, Kulla HG, Leisinger T (1982) Properties of purified Orange II azoreductase, the enzyme initiating azo dye degradation by *Pseudomonas* KF46. Eur J Biochem 129:197–203
17. Zimmermann T, Gasser F, Kulla HG, Leisinger T (1984) Comparison of two azoreductases acquarid during adaptation to growth on azo dyes. Arch Microbiol 138:37–43
18. Moutaouakkil A, Zeroual Y, Zzayri FZ, Talbi M, Lee K, Blaghen M (2003) Purification and partial characterization of azoreductase from *Enterobacter agglomerans*. Arch Biochem Biophys 413:139–146
19. Chen H, Wang RF, Cerniglia CE (2004) Molecular cloning, overexpression, purification, and characterization of an aerobic FMN-dependent azoreductase from *Enterococcus faecalis*. Protein Expr Purif 34:302–310
20. Chen H, Hopper S, Cerniglia CE (2005) Biochemical and molecular characterization of an azoreductase from *Staphylococcus aureus*, a tetrameric NADPH-dependent flavoprotein. Microbiology 151:1433–1441
21. Nakanishi M, Yatome C, Ishida N, Kitade Y (2001) Putative ACP phosphodiesterase gene (acpD) encodes an azoreductase. J Biol Chem 276:46394–46399
22. Suzuki T, Timofei S, Kurunczi L, Dietze U, Schuurmann G (2001) Correlation of aerobic biodegradability of sulfonated azo dyes with the chemical structure. Chemosphere 45:1–9
23. Yan B, Zhou J, Wang J, Du C, Hou H, Song Z, Bao Y (2004) Expression and characteristics of the gene encoding azoreductase from *Rhodobacter sphaeroides* AS1.1737. FEMS Microbiol Lett 236:129–136
24. Liger D, Graille M, Zhou CZ, Leulliot N, Quevillon-Cheruel S, Blondeau K, Janin J, van Tilbeurgh H (2004) Crystal structure and functional characterization of yeast YLR011wp, an enzyme with NAD(P)H-FMN and ferric iron reductase activities. J Biol Chem 279:34890–34897
25. Ramalho PA, Cardoso MH, Cavaco-Paulo A, Ramalho MT (2004) Characterization of azo reduction activity in a novel ascomycete yeast strain. Appl Environ Microbiol 70:2279–2288
26. Carvalho MC, Pereira C, Goncalves IC, Pinheiro HM, Santos AR, Lopes A, Ferra MI (2008) Assessment of the biodegradability of a monosulfonated azo dye and aromatic amines. Int Biodeterior Biodegradation 62(2):96–103
27. Pearce C, Guthrie JT, Lloyd JR (2008) Reduction of pigment dispersions by *Shewanella* strain J18 143. Dyes Pigm 76(3):696–705
28. Mechsner K, Wuhrmann K (1982) Cell permeability as a rate limiting factor in the microbial reduction of sulfonated azo dyes. Eur J Appl Microbiol Biotechnol 15(2):123–126
29. Chung K-T, Stevens SE (1993) Degradation of azo dyes by environmental microorganisms and helminths. Environ Toxicol Chem 12:2121–2132
30. Brown MA, De Vito SC (1993) Predicting azo dye toxicity. Crit Rev Environ Sci Technol 23:249–324

31. Zbaida S (1995) The mechanisms of microsomal azoreduction: predictions based on electronic aspects of structure-activity relationship. Drug metab Rev 27:497–516
32. Coughlin MF, Kinkle BK, Tepper A, Bishop PL (1997) Characterization of aerobic azo dye-degrading bacteria and their activity in biofilms. Water Sci Tech 36:215–220
33. Coughlin MF, Kinkle BK, Bishop PL (1999) Degradation of azo dyes containing aminonaphthol by *Sphingomonas* sp strain 1CX. Ind Microbiol Biotechnol 23:341–346
34. Dykes GA, Timm RG, von Holy A (1994) Azoreductase activity in bacteria associated with the greening of instant chocolate puddings. Appl Environ Microbiol 60:3027–3029
35. Jiang H, Bishop PL (1994) Aerobic biodegradation of azo dyes in biofilms. Water Sci Tech 29:525–530
36. Sugiura W, Miyashita T, Yokoyama T, Arai M (1999) Isolation of azo-dye degrading microorganisms and their application to white discharge printing of fabric. J Biosci Bioeng 88:577–581
37. Goszczynski S, Paszczynski A, Pasti-Grisbi MB, Crowford RL, Crawford DL (1994) New pathway for degradation of sulfonated azo dyes by microbial peroxidases of *Phanerochaete chrysosporium* and *Streptomyces chromofuscus*. J Bacteriol 176:1339–1347
38. Munari FM, Gaio TA, Calloni R, Dillon AJP (2008) Decolorization of textile dyes by enzymatic extract and submerged cultures of *Pleurotus sajor-caju*. World J Microbiol Biotechnol 24:1383–1392
39. Southern TG (1995) Technical solutions to the colour problem: a critical review. In: Cooper P (ed) Colour in dyehouse effluent. Society of Dyes and Colourists, Bradford, p 75
40. Bras R, Ferra IA, Pinheiro HM, Goncalves IC (2001) Batch tests for assessing decolorization of azo dyes by methanogenic and mixed cultures. J Biotechnol 89:155–162
41. El Ahwany AMD (2008) Decolorization of Fast red by metabolizing cells of *Oenococcus oeni* ML34. World J Microbiol Biotechnol 24:1521–1527
42. Wong PK, Yuen PY (1996) Decolorization and biodegradation of methyl red by *Klebsiella pneumoniae* RS-13. Water Res 30:1736–1744
43. Steffan S, Bardi L, Marzona M (2005) Azo dyes biodegradation by microbial cultures immobilized in alginate beads. Environ Int Special Issue: Recent Adv Bioremediat 31 (2):201–205
44. Ozdemir G, Pazarbasi B, Kocyigit A, Omeroglu EE, Yasa I, Karaboz I (2008) Decolorization of Acid Black 210 by *Vibrio harveyi* TEMS1, a newly isolated bioluminescent bacterium from Izmir Bay, Turkey. World J Microbiol Biotechnol 24:1375–1381
45. Khalid A, Arshad M, Crowley DE (2008) Accelerated decolorization of structurally different azo dyes by newly isolated bacterial strains. Appl Microbiol Biotechnol 78:361–369
46. Maximo C, Costa-Ferreira MC (2004) Decolourisation of reactive textile dyes by *Irpex lacteus* and lignin modifying enzymes. Process Biochem 39:1475–1479
47. Tatarko M, Bumpus JA (1998) Biodegradation of Congo Red by *Phanerochaete chrysosporium*. Water Res 32:1713–1717
48. Wang Y, Yu J (1998) Adsorption and degradation of synthetic dyes on the mycelium of *Trametes versicolor*. Water Sci Technol 38:233–238
49. Yesiladalil SK, Pekin GI, Bermek H, Arslan-Alaton I, Orhon D, Tamerler C (2006) Bioremediation of textile azo dyes by *Trichophyton rubrum* LSK-27. World J Microbiol Biotechnol 22:1027–1031
50. Mohan SV, Ramanajah SV, Sarma PN (2008) Biosorption of direct azo dye from aqueous phase onto *Spirogyra* sp. IO2: evaluation of kinetics and mechanistic aspects. Biochem Eng J 38(1):61–69
51. Kapdan KI, Kargi F, Mullan G, Marchant R (2000) Decolorization of textile dyestuff by a mixed bacterial consortium. Biotechnol Lett 22:1179–1181
52. Pearce CI, Lioyd JR, Guthrie JT (2003) The removal of colour from textile wastewater using whole bacterial cells: a review. Dye Pigm 58:179–196
53. Yatman HC, Akyol A, Bayramoglu M (2004) Kinetics of photocatalytic decolorization of an azo reactive dye in aqueous ZnO suspensions. Ind Eng Chem Res 43:6035–6039

54. Khalid A, Arshad M, Crowley DE (2008) Decolorization of azo dyes by *Shewanella* sp. under saline conditions. Appl Microbiol Biotechnol 79:1053–1059
55. Mohanty S, Dafale N, Rao NN (2006) Microbial decolorization of reactive Black-5 in a two-stage anaerobic-aerobic reactor using acclimatized activated textile sludge. Biodegradation 17:403–413
56. Yang Q, Tao L, Yang M, Zhang H (2008) Effects of glucose on the decolorization of Reactive Black 5 by yeast isolates. J Environ Sci 20(1):105–108
57. Kim SY, An JY, Kim BW (2008) The effects of reductant and carbon source on the microbial decolorization of azo dyes in an anaerobic sludge process. Dye Pigm 76(1):256–263
58. Kulla HG (1981) Aerobic bacterial degradation of azo dyes. In: Leisinger T, Cook AM, Nüesch J, Hütter R (eds) Microbial degradation of xenobiotics and recalcitrant compounds. Academic, London, pp 387–399
59. Kulla HG, Krieg R, Zimmermann T, Leisinger T (1984) Experimental evolution of azo dye-degrading bacteria. In: Klug MJ, Reddy CA (eds) Current perspectives in microbial ecology. American Society of Microbiology, Washington, DC, pp 663–667
60. Kulla HG, Klausener F, Meyer U, Lüdeke B, Leisinger T (1983) Interference of aromatic sulfo groups in the microbial degradation of the azo dyes Orange I and Orange II. Arch Microbiol 135:1–7
61. Blümel S, Contzen M, Lutz M, Stolz A, Knackmuss H-J (1998) Isolation of a bacterial strain with the ability to utilize the sulfonated azo compound 4-carboxy-4′-sulfoazobenzene as sole source of carbon and energy. Appl Environ Microbiol 64:2315–2317
62. Feigel BJ, Knackmuss H-J (1993) Syntrophic interactions during degradation of 4-aminobenzenesulfonic acid by a two species bacterial culture. Arch Microbiol 159:124–130
63. Carliell CM, Barcaly SJ, Shaw C, Wheatly AD, Buckley CA (1998) The effect of salts used in textile dyeing on microbial decolorization of a reactive azo dye. Environ Technol 19:1133–1137
64. Lourenco ND, Novais JM, Pinheiro HM (2000) Reactive textile dye colour removal in a sequencing batch reactor. Water Sci Technol 42:321–328
65. Chang JS, Chou C, Lin YC, Lin PJ, Ho JY, Hu TL (2001) Kinetics characteristics of bacterial azo dye decolorization by *Pseudomonas luteola*. Water Res 35:2841–2850
66. Hu TL (1996) Removal of reactive dyes from aqueous solutions by different bacterial genera. Water Sci Technol 34:89–95
67. Junnarkar N, Murty JD, Bhatt NS, Madamwar D (2006) Decolorization of diazo dye Direct Red 81 by a novel bacterial consortium. World J Microbiol Biotechnol 22:163–168
68. Susla M, Svobodova K (2008) Effect of various synthetic dyes on the production of manganese-dependent peroxidase isoenzymes by immobilized *Irpex lacteus*. World J Microbiol Biotechnol 24:225–230
69. Perez J, De La Rubia T, Ben Hamman O, And Martinez J (1998) *Phanerochaete flavido-alba* laccase induction and modification of manganese peroxidase isoenzyme pattern in decolorized olive oil mill wastewaters. Appl Environ Microbiol 64(7):2726–2729
70. Camarero S, Ibarra D, Martinez MJ, Martinez AT (2005) Lignin-derived compounds as efficient laccase mediators for decolorization of different types of recalcitrant dyes. Appl Environ Microbiol 71(4):1775–1784
71. Gettemy JM, Ma B, Alic M, Gold MH (1998) Reverse transcription PCR analysis of the regulation on the manganese peroxidases gene family. Appl Environ Microbiol 64(2):569–574
72. Moreira MH, Feijoo G, Palma C, Lema JM (1997) Continuous production of manganese peroxidase by *Phanerochaete chrysosporium* immobilized on polyurethane foam in pulsed packed bed bioreactor. Biotechnol Bioeng 56:130–137
73. Rogalski J, Szczodrak J, Janusz G (2006) Manganese peroxidases production in submerged cultures by free and immobilized mycelia of *Nematoloma frowardii*. Bioresour Technol 97:469–476
74. Martinez AT (2002) Molecular biology and structure-function of lignin-degrading heme peroxidases. Enzyme Microb Technol 30:425–444

75. Adedayo O, Javadpour S, Taylor C, Anderson WA, Moo-Young M (2004) Decolourization and detoxification of Methyl Red by aerobic bacteria from a wastewater treatment plant. World J Microbiol Biotechnol 20:545–550
76. Angelova B, Avramova T, Stefanova L, Mutafov S (2008) Temperature effect of bacterial azo bond reduction kinetics: an Arrhenius plot analysis. Biodegradation 19(3):387–393
77. Chang J, Kuo T (2000) Kinetics of bacterial decolorization of azo dye with *Escherichia coli* NO3. Bioresour Technol 75:107–111
78. Mali PL, Mahajan MM, Patil DP, Kulkarni MV (2000) Biodecolorization of members of triphenylmethanes and azo groups of dyes. J Sci Ind Res India 59:221–224
79. Asad S, Amoozegar MA, Pourbabaee AA, Sarbolouki MN, Dastgheib SMM (2007) Decolorization of textile azo dyes by newly isolated halophilic and halotolerant bacteria. Bioresour Technol 98:2082–2088
80. Uddin SM, Zhou J, Qu Y, Wang P, Zhao LH (2007) Biodecolorization of azo dye Acid Red B under high salinity condition. Bull Environ Contam Toxicol 79:440–444

Index